Leckie × Leckie
Scotland's leading educational publishers

1 FOR REVISION

National 5 CHEMISTRY SUCCESS GUIDE

N5 CHEMISTRY SUCCESS GUIDE

Bob Wilson

001/27092013

10 9 8 7 6 5 4 3

ISBN 9780007504695

Published by
Leckie & Leckie Ltd
An imprint of HarperCollins*Publishers*
Westerhill Road, Bishopbriggs, Glasgow, G64 2QT
T: 0844 576 8126 F: 0844 576 8131
leckieandleckie@harpercollins.co.uk www.leckieandleckie.co.uk

Special thanks to
Ink Tank (cover design); QBS (layout and illustration); Jill Laidlaw (copy editing); Paul Sensecall (proofreading)

A CIP Catalogue record for this book is available from the British Library.

Acknowledgements
We would like to thank the following for permission to reproduce their material:
p. 16 POWER AND SYRED/SCIENCE PHOTO LIBRARY; p. 49 LIFE IN VIEW/SCIENCE PHOTO LIBRARY; p.61, 93 ANDREW LAMBERT PHOTOGRAPHY/SCIENCE PHOTO LIBRARY; p. 62 E. R. DEGGINGER/SCIENCE PHOTO LIBRARY; p. 87 JAMES KING-HOLMES/SCIENCE PHOTO LIBRARY; p. 92 MARTYN F. CHILLMAID/SCIENCE PHOTO LIBRARY; p. 95 NASA/GSFC/ SCIENCE PHOTO LIBRARY

Whilst every effort has been made to trace the copyright holders, in cases where this has been unsuccessful, or if any have inadvertently been overlooked, the Publishers would gladly receive any information enabling them to rectify any error or omission at the first opportunity.

Printed in Italy by Grafica Veneta S.P.A.

Contents

Unit 1: Chemical changes and structure

Unit 2: Nature's chemistry

Unit 3: Chemistry in society

Rates of reaction

Remember from National 4!

The rate of a reaction can be increased by: increasing the concentration; decreasing the particle size; raising the temperature and adding a catalyst.

The change in rate of a chemical reaction can be followed by measuring the volume of gas produced and the change in mass of reactants over time.

Calculating average rate of reaction

The **average rate** of a chemical reaction is the change in the quantity of reactant or product over time and can be calculated using data collected from experiments carried out in the laboratory.

A suitable reaction is marble chips (containing calcium carbonate) reacting with hydrochloric acid:

calcium carbonate + hydrochloric acid → calcium chloride + water + carbon dioxide

$$CaCO_3(s) + 2\,HCl(aq) \rightarrow CaCl_2(aq) + H_2O(\ell) + CO_2(g)$$

Measuring the change in volume of gas produced during the reaction

The rate at which carbon dioxide gas is given off can be measured by collecting the gas and measuring the volume at fixed time intervals. Collecting the gas by displacement of water is the most commonly used method in the laboratory, but using a syringe gives more accurate measurements.

The graph shows how the volume of gas changes over the course of the reaction.

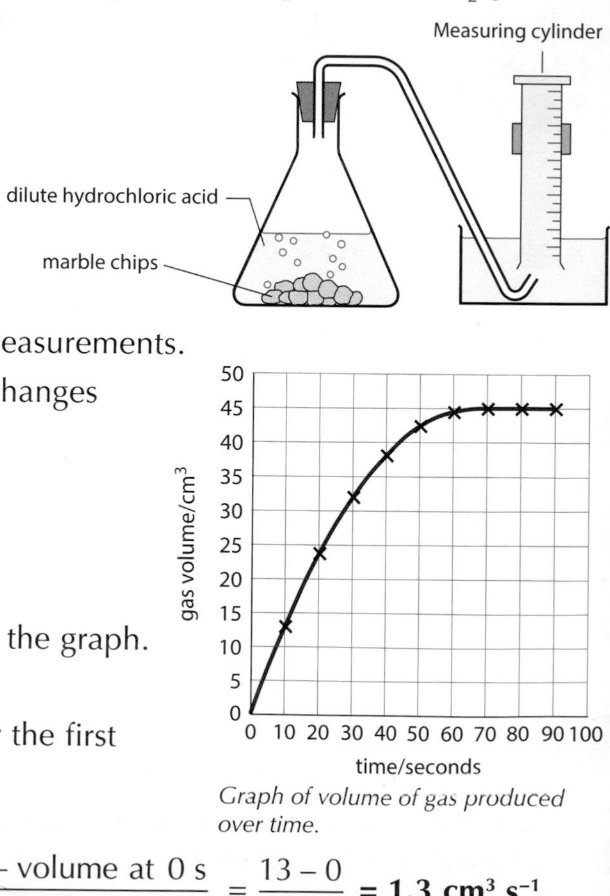

Graph of volume of gas produced over time.

Remember!

$$\textbf{Average rate} = \frac{\textbf{change in volume}}{\textbf{change in time}}$$

The information required is obtained from the graph.

Example 1

Calculate the average rate of reaction over the first 10 seconds.

Worked answer:

$$\text{Average rate of reaction} = \frac{\text{volume at 10 s} - \text{volume at 0 s}}{\text{time interval}} = \frac{13 - 0}{10 - 0} = \mathbf{1.3\ cm^3\ s^{-1}}$$

Since rate here is a measure of the change in volume of the gas over time, the unit of rate is **$cm^3\ s^{-1}$** (cubic centimetres per second).

Measuring the loss in mass during a reaction

The rate at which carbon dioxide is given off is obtained by measuring the loss in mass of the reactants at regular time intervals. As the gas is produced, it is released into the air so the mass of the reaction mixture and flask decreases.

The graph shows how the mass changes over the course of the reaction.

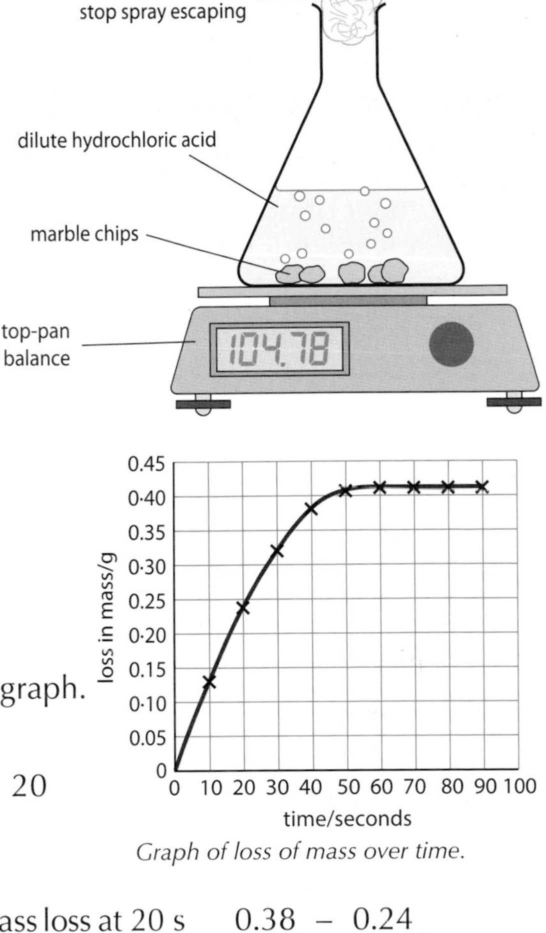

Graph of loss of mass over time.

Remember!

$$\textbf{Average rate} = \frac{\textbf{change in mass}}{\textbf{change in time}}$$

The information required is obtained from the graph.

Example 2

Calculate the average rate of reaction between 20 and 40 seconds.

Worked answer:

$$\text{Average rate of reaction} = \frac{\text{mass loss at 40 s} - \text{mass loss at 20 s}}{\text{time interval}} = \frac{0.38 - 0.24}{40 - 20}$$

Average rate of reaction = 0.007 g s^{-1}

Since rate here is a measure of the mass loss over time, the unit of rate is **g s^{-1}** (grams per second).

Quick Test

1. Look at the graph of volume of gas against time. Calculate the average rate of reaction between 30 and 50 seconds.
2. Look at the graph of loss of mass against time. Calculate the average rate of reaction over the first 20 seconds.

Nuclide notation, isotopes and relative atomic mass (RAM)

Remember from National 4!

The number of protons (p) in an atom is called the atomic number. The mass number is the number of protons and neutrons (n) added together.

Nuclide notation

Nuclide notation is a shorthand way of showing the mass number and atomic number of an **atom** along with the symbol of the element. The nuclide notation for an atom of chlorine, atomic number 17, with a mass number of 35, can be used as an example:

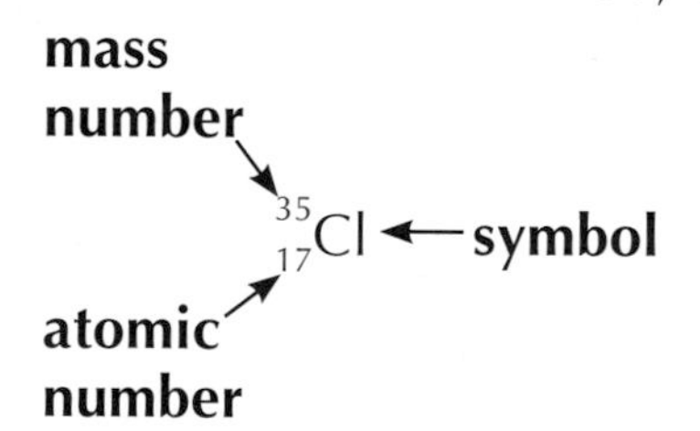

Nuclide notation can also be used for **ions**. The number of electrons (e^-) is the only thing that changes when an atom forms an ion. This means the atomic number and the mass number do not change. The nuclide notation for an atom and ion of chlorine is shown below.

Atom

$^{35}_{17}Cl$: p = 17, e^- = 17 } overall zero charge

n = 35 – 17 = 18

Ion

$^{35}_{17}Cl^-$: p = 17, e^- = 18 } overall 1– charge

n = 35 – 17 = 18

Isotopes

Atoms that have the same atomic number but different mass numbers are known as **isotopes**. Some atoms of chlorine, for example, have 18 neutrons, while others have 20 neutrons. This means their mass numbers are different (35 and 37) but their atomic numbers are the same (17). The nuclide notation for the two isotopes of chlorine is: $^{35}_{17}Cl$ and $^{37}_{17}Cl$.

Remember!

The isotopes of an element have the same electron arrangement so they have identical chemical behaviour.

Relative Atomic Mass (RAM)

TOP TIP

A table of RAM values for some common elements can be found in the SQA data booklet.

The total mass of an atom comes from the mass of its neutrons and protons. Most elements however have two or more isotopes so an average is taken of the mass of all the isotopes. This average mass is called the **relative atomic mass (RAM).**

We get the information needed to calculate the relative atomic mass from a mass spectrometer. The information is often given in a graph.

From the graph we can see that:

- Chlorine has two isotopes – corresponding to the two peaks.
- The isotopes have atoms of mass 35 and 37.
- 75% of atoms have a mass of 35, and 25% have a mass of 37.

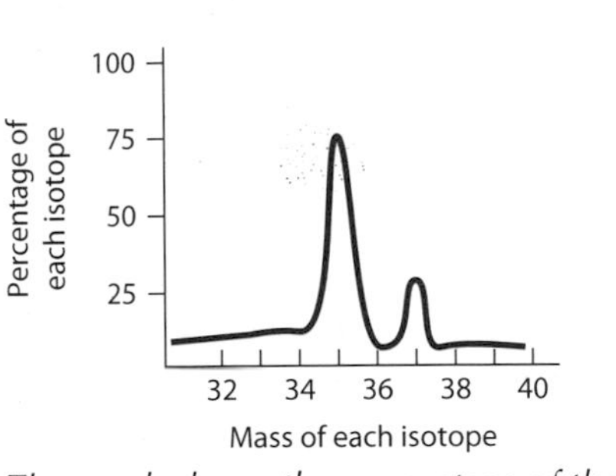

The graph shows the percentage of the two isotopes of chlorine.

This information can be used to calculate the relative atomic mass of chlorine:

$$\text{RAM} = \frac{(75 \times 35) + (25 \times 37)}{100} = \mathbf{35.5}$$

Quick Test

1. An atom of bromine has 44 neutrons and 35 protons.
   (a) Write the nuclide notation for the atom.
   (b) When the atom forms an ion, it gains an electron. Write the nuclide notation for the bromide ion.
   (c) Bromine has two stable isotopes, ${}^{79}_{35}Br$ and ${}^{81}_{35}Br$. The relative atomic mass of bromine is sometimes rounded up to 80. What can you deduce about the percentage abundance of each isotope?
2. The table gives information about the three naturally occurring isotopes of neon.

Isotope	Percentage abundance
${}^{20}Ne$	90.5
${}^{21}Ne$	0.3
${}^{22}Ne$	9.2

Calculate the relative atomic mass of neon.

Covalent bonding and shapes of molecules

Remember from National 4!

Atoms of non-metal elements achieve a stable electron arrangement by sharing outer electrons and forming a covalent bond.

Single covalent bonds

The two fluorine atoms in a fluorine molecule are held together by a **single** covalent bond – a **shared pair of electrons**. This can be shown in a dot and cross diagram.

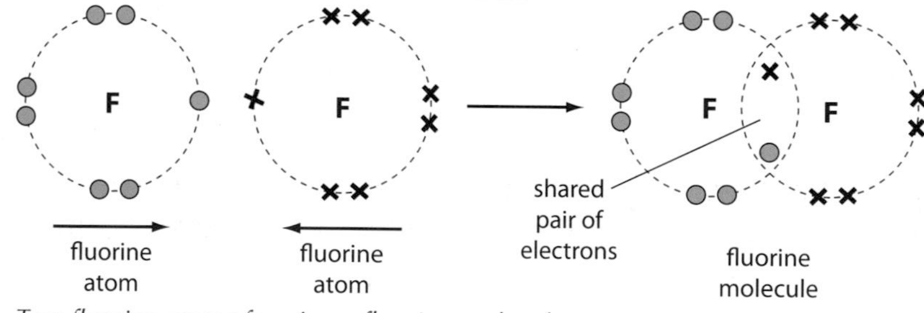

Two fluorine atoms forming a fluorine molecule.

In a covalent bond, the positive nucleus of each atom attracts not only its own electrons but also the electrons from the other atom.

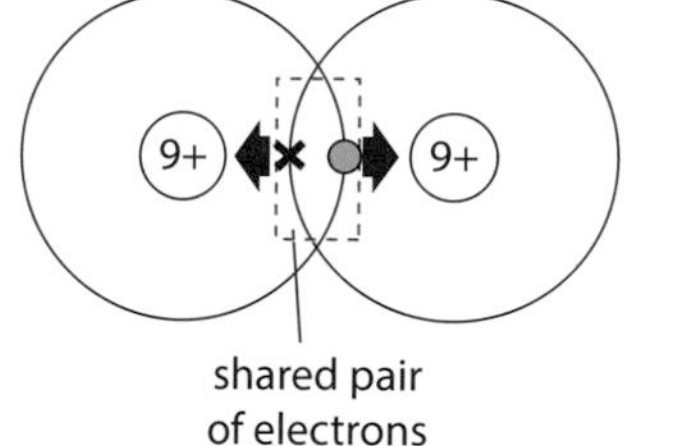

The nucleus of one atom attracts the electrons of the other.

Count the electrons in the outer energy level of each atom in the fluorine molecule – each has seven electrons of its own and a share in an eighth. Eight electrons is a stable arrangement.

A hydrogen atom and a chlorine atom join to form a hydrogen chloride molecule – only the outer electrons are shown for the chlorine atom:

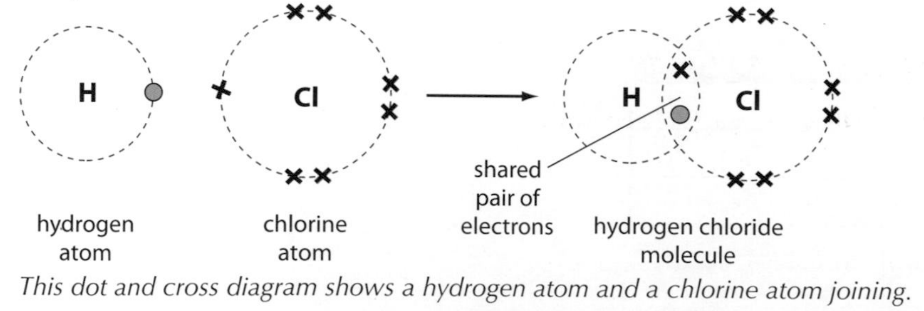

This dot and cross diagram shows a hydrogen atom and a chlorine atom joining.

Count the number of electrons in the outer energy level of each atom in the molecule. Chlorine has eight electrons, the stable arrangement. Hydrogen has two electrons, which at first glance is not a stable arrangement, but it is the same arrangement as the noble gas helium – two electrons in the first (only) energy level is a stable arrangement.

Fluorine and hydrogen chloride are **diatomic molecules** – fluorine is an element and hydrogen chloride a compound. They have a **linear** shape – their atoms are in a line. A fluorine molecule can be written as F–F and hydrogen chloride as H–Cl. This representation shows both the covalent bond and the shape of the molecule.

Hydrogen, H_2

Sharing their electron gives each hydrogen atom a filled outer energy level. The hydrogen molecule is diatomic.

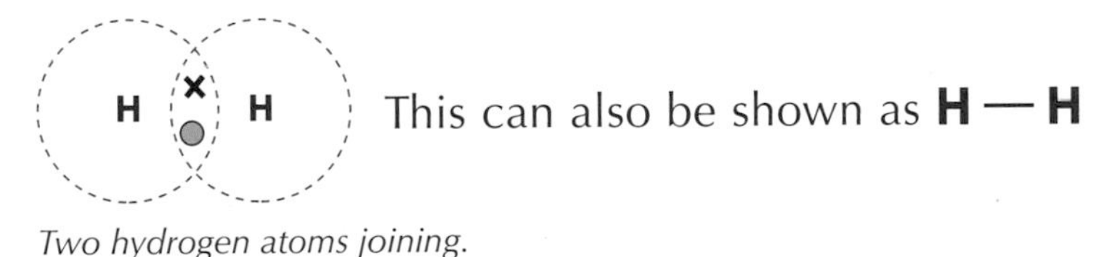

Two hydrogen atoms joining.

Water, H_2O

Atoms of oxygen and hydrogen combine by sharing pairs of electrons to form water. The oxygen atom has six outer electrons so needs two more. Oxygen forms two single covalent bonds to hydrogen atoms:

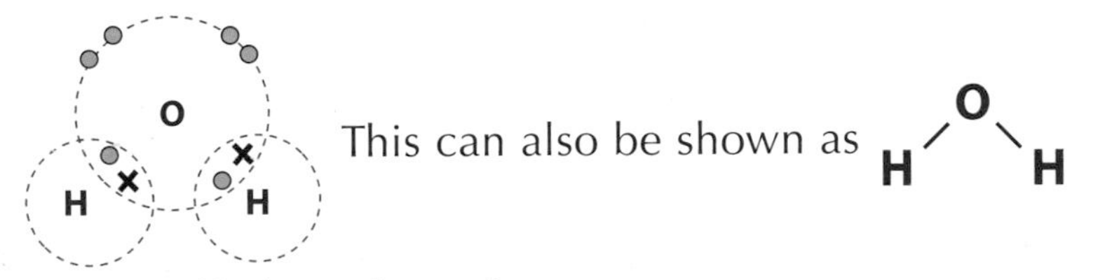

Oxygen and hydrogen sharing electrons.

The shape of a water molecule is **bent**.

Ammonia, NH_3

Atoms of nitrogen and hydrogen combine by sharing pairs of electrons to form ammonia. The nitrogen atom has five outer electrons so needs three more. Nitrogen forms three single covalent bonds to hydrogen atoms:

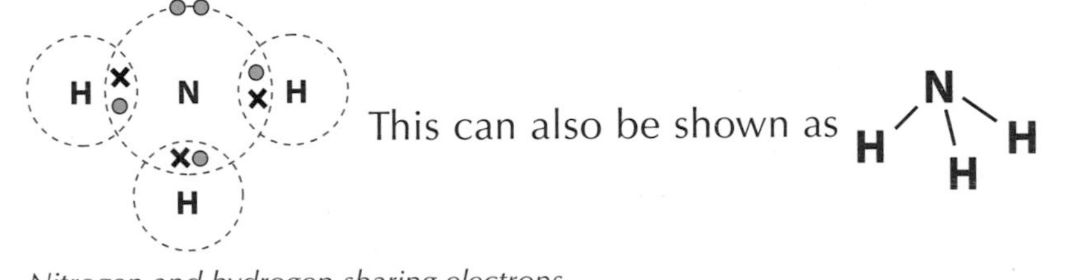

Nitrogen and hydrogen sharing electrons.

The shape of an ammonia molecule is **pyramidal**.

Methane, CH_4

Atoms of carbon and hydrogen combine by sharing pairs of electrons to form methane. The carbon atom has four outer electrons so needs four more. Carbon forms four single covalent bonds to hydrogen atoms:

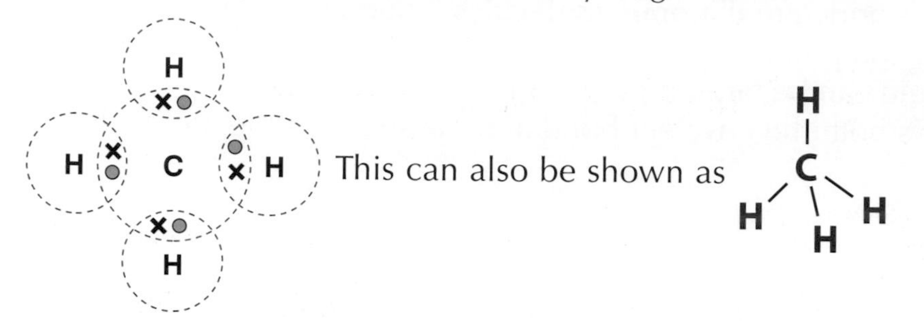

Carbon and hydrogen sharing electrons.

TOP TIP

The table gives a general rule about the number of atoms in a molecule with single bonds and the shape of the molecule.

The shape of a methane molecule is **tetrahedral**.

Number of atoms in molecule	Shape of molecule
two	linear
three	bent
four	pyramidal
five	tetrahedral

Numbers of atoms and the shapes of molecules.

Double and triple covalent bonds

It is possible for the atoms of non-metal elements to join together in such a way as to form **double** and even **triple** covalent bonds.

Oxygen, O_2

Both oxygen atoms have six outer electrons so both need two more electrons to form a stable arrangement of electrons (eight). They do this by sharing two pairs of electrons forming a double covalent bond:

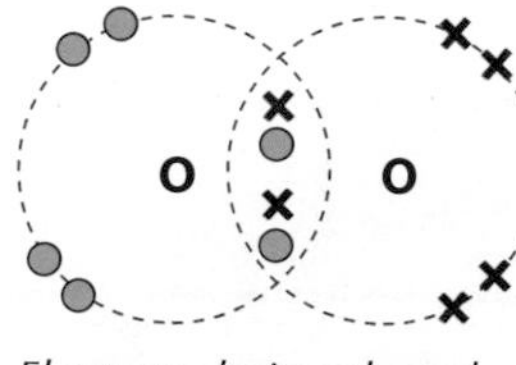

This can also be shown as **O = O**

Electrons being shared by two oxygen atoms.

Nitrogen, N_2

Both nitrogen atoms have five outer electrons so both need three more electrons to form a stable arrangement of electrons. They do this by sharing three pairs of electrons forming a triple covalent bond:

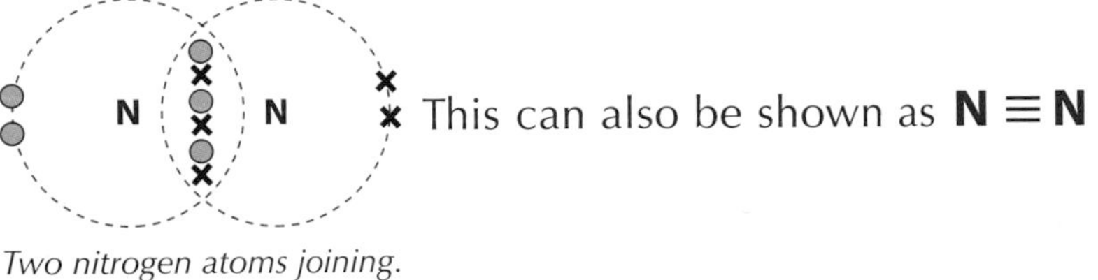

This can also be shown as **N≡N**

Two nitrogen atoms joining.

Carbon dioxide, CO_2

Carbon has four outer electrons so needs four more. It forms double covalent bonds with two oxygen atoms so that all the atoms now have a full outer shell of electrons:

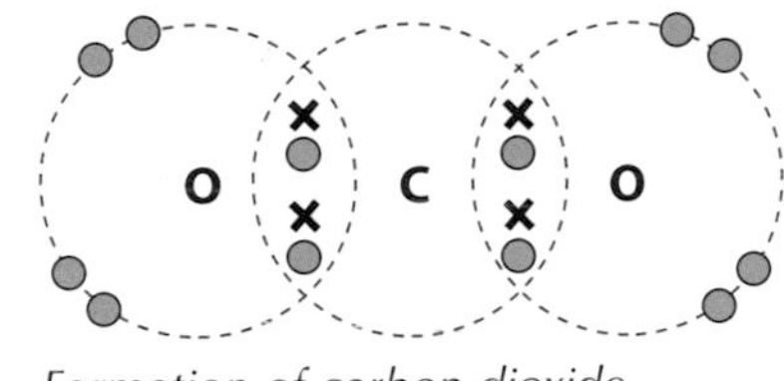

This can also be shown as **O=C=O**

Formation of carbon dioxide.

Note that the shape of the carbon dioxide molecule is linear and not bent as might be expected, because of the double bonds.

Quick Test

1. Hydrogen selenide is a flammable gas formed between hydrogen and selenium.
   (a) Draw a dot and cross diagram to show how hydrogen and selenium atoms join to form a molecule.
   (b) Draw the likely shape of hydrogen selenide.
   (c) Write the molecular formula for hydrogen selenide.
2. (a) Draw a dot and cross diagram to show how phosphorus and hydrogen atoms join to form a molecule.
   (b) Draw the likely shape of the molecule formed in (a).
   (c) Write the molecular formula for the compound.
3. Silane is formed when hydrogen reacts with silicon.
   (a) Draw a dot and cross diagram to show how hydrogen and silicon atoms join to form a molecule.
   (b) Draw the likely shape of silane.
   (c) Write the molecular formula for silane.
4. (a) Use dot and cross diagrams to show how carbon and sulfur atoms join to form carbon disulfide (CS_2).
   (b) Draw the likely shape of the carbon disulfide molecule.
   (Hint: look at how carbon dioxide is formed.)

Structure and properties of covalent substances

Remember from National 4!

Covalent substances that are made up of **individual molecules** can exist as **gases, liquids or solids** at room temperature.

Covalent molecular

The table shows the melting and boiling points of some covalent molecular substances and their states at room temperature. Note how the melting and boiling points increase as you move across from gases to liquids to solids.

gases	liquids	solids
O = O O_2: mp = –218 °C bp = –183 °C	Br — Br Br_2: mp = –7 °C bp = 59 °C	P_4 (tetrahedron of four P atoms) P_4: mp = 44 °C bp = 280 °C
CH_4 (C bonded to four H) CH_4: mp = –182.5 °C bp = –164 °C	H–O–H H_2O: mp = 0 °C bp = 100 °C	S_8 (ring of eight S atoms) S_8: mp = 113 °C bp = 445 °C

Covalent gases, liquids and solids

The covalent bonds holding the atoms together are strong, but the **forces between the molecules are weak,** and it doesn't take a lot of energy to separate the molecules. It is these weak forces between molecules that have to be broken when a molecular substance melts or boils. This results in **molecular substances** generally having **low melting and boiling points**.

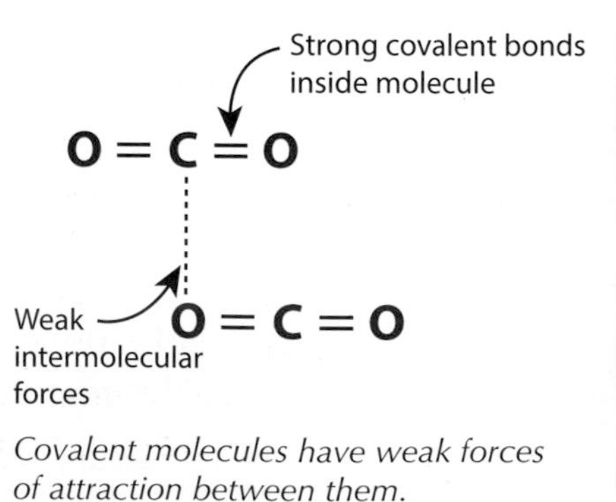

Covalent molecules have weak forces of attraction between them.

Covalent network

Some covalently bonded substances form giant three-dimensional structures called **covalent networks** in which all the atoms are covalently bonded to each other. There are no individual molecules. Covalent bonds are relatively strong so it takes a lot of energy to break a covalent bond. Network substances therefore have **very high melting and boiling points**. Diamond and silicon dioxide are common examples of covalent networks.

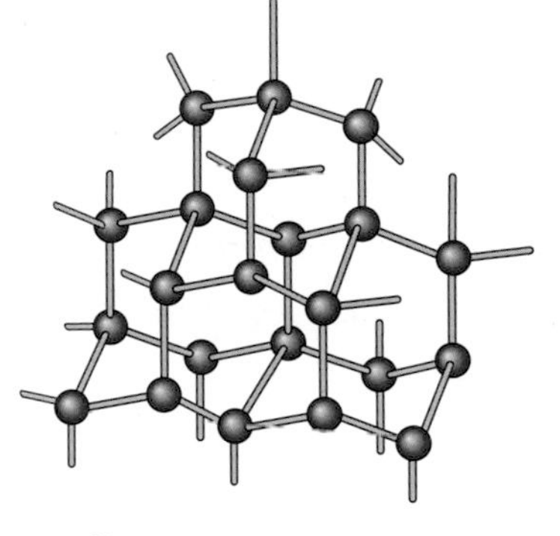

Carbon atom

***Part** of diamond's covalent network.*

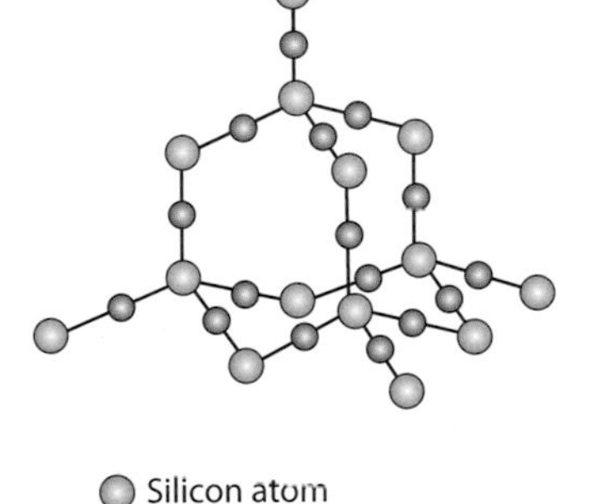

Silicon atom

Oxygen atom

***Part** of silicon dioxide's covalent network.*

The three-dimensional network makes diamond and silicon dioxide extremely strong – diamond is the hardest known substance. **Covalent network substances** are generally **very strong**. Covalent substances are generally **insoluble in water** but can be soluble in covalent liquids, such as white spirit. Covalent liquids tend to evaporate quickly and can be very flammable. Covalent substances (with the exception of carbon in the form of graphite) **do not conduct electricity** in any state because they have no electrons free to move from atom to atom.

Quick Test

1. Complete the summary. You may wish to use the word bank to help you.

Covalent substances exist as either individual (a)__________ or giant (b)__________. Many molecular substances are (c)_______or liquids. This is because although the (d)__________ bonds that hold the atoms together are relatively (e)_______, the forces between the molecules are (f)________. It does not take a lot of (g)__________ to separate the (h)__________. This results in molecular substances having (i)_________ melting and boiling points. Covalent network substances have (j)_________ melting and boiling points. This is because each atom in the network is (k)__________ bonded to other atoms resulting in a very strong (l)_________ structure, which needs a lot of energy to break the (m)_________. Covalent substances are generally (n)___________ in water but can (o)__________ in covalent liquids. Covalent substances do not conduct (p)___________ because the (q)_____ are not free to move from atom to atom.

Word bank

three-dimensional, bonds, covalent, covalently, dissolve, electricity, electrons, energy, gases, high, insoluble, low, molecules, networks, separate, strong, weak

Structure and properties of ionic compounds

Remember from National 4!

Ionic compounds form when metal atoms transfer electrons to non-metal atoms resulting in positive metal ions and negative non-metal ions. The attraction of positive and negative ions is called an ionic bond.

Structure of ionic compounds

In ionic compounds the oppositely charged positive (metal) ions and negative (non-metal) ions form a giant structure known as an **ionic lattice**. Sodium chloride (NaCl) is a good example of an ionic lattice.

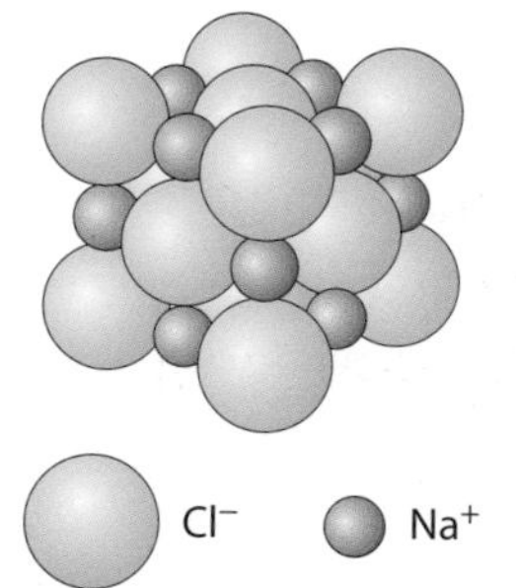

The ions in sodium chloride form a three-dimensional lattice.

Sodium chloride crystals are cubic shaped.

There are strong **electrostatic forces** of attraction between the oppositely charged ions. This results in each ion being surrounded by ions carrying the opposite charge. A giant three-dimensional structure (ionic lattice) is produced.

Properties

Ionic compounds are solids at room temperature and have many properties in common. These properties include:

- **Electrical non-conductors as solids** – the ions are not free to move in the lattice.
- **Electrical conductors when in solution or a molten liquid** – the ions are free to move and act as charge carriers.
- **High melting and boiling points** – it takes a lot of energy to break the ions apart.
- **Solubility in water** – the regular pattern of the ions in the lattice is broken down, and the ions are surrounded by water molecules.

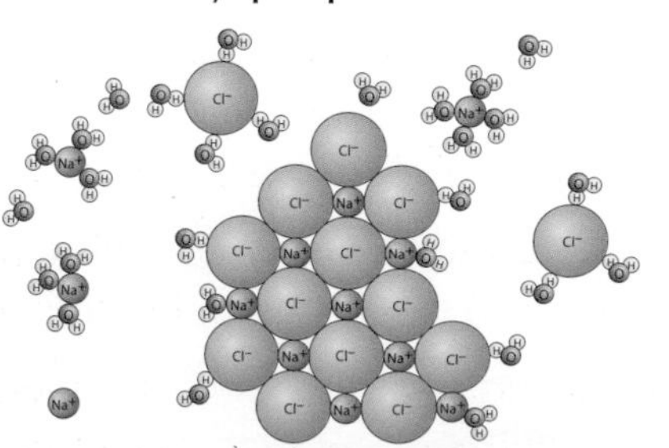

The sodium chloride crystal lattice breaks down when it dissolves in water.

Comparing ionic and covalent properties

Colour

The colour of the substance can often indicate the type of bonding – many **ionic** substances are highly coloured.

The blue colour in copper sulphate crystals and the red in the gem stone ruby indicate the presence of ions.

State

If a substance is a **gas or liquid** at room temperature then the substance will be **covalently** bonded and exist as individual molecules. **Solids** are more difficult to distinguish from each other. If a solid can be **easily melted**, then it will be **covalent molecular**. Wax (a hydrocarbon) melts quickly in boiling water, indicating that it is covalent molecular. The liquid wax floats on the surface of the water and doesn't dissolve, which indicates that it is covalent not ionic

Conductivity

Wax doesn't conduct electricity either as a solid or as a liquid, verifying that it is not ionic. The same test carried out on an ionic compound like sodium chloride, when it is in the liquid state, would show that it conducts, indicating ionic bonding.

Quick Test

1. The table shows some of the properties of potassium fluoride.

Property	
Melting point /°C	857
Boiling point /°C	1502
Electrical conductor as a solid?	no
Electrical conductor as a liquid or in solution?	yes

Suggest the type of bonding present in potassium fluoride and give two pieces of evidence from the table to support your answer.

2. A student put some solid magnesium chloride into a beaker and tested its conductivity. He concluded that magnesium chloride was covalent because it did not conduct electricity. Using your knowledge of chemistry, comment on his conclusion.

3. Sodium chloride is a solid with a melting point of 801°C, but hydrogen chloride is a gas at room temperature. Explain these observations.

Chemical formulae using group ions

Remember from National 4!

It is important that you remember the rules for working out formulae of elements and compounds containing two elements and writing equations because they can be used to write formulae and equations involving more complicated compounds.

Formulae of elements

Most elements have their symbol as their formula, e.g. potassium: K; magnesium: Mg; sulfur: S. The exceptions are the diatomic molecules: H_2, N_2, O_2, F_2, Cl_2, Br_2 and I_2.

Formulae of compounds using valency

Valency is another word for combining power. The valency of an element tells us how many bonds it can form with another atom. Valencies of the main group elements depend on their position in the periodic table.

Group 1	Group 2	Group 3	Group 4	Group 5	Group 6	Group 7	Group 0
H							He
Li	Be	B	C	N	O	F	Ne
Na	Mg	Al	Si	P	S	Cl	Ar
K	Ca	Ga	Ge	As	Se	Br	Kr
Valency 1	**Valency 2**	**Valency 3**	**Valency 4**	**Valency 3**	**Valency 2**	**Valency 1**	**Valency 0**

The 'cross over' method is the simplest way to work out formulae.

- Write the symbols for the elements.
- Work out the valency and write it under the symbols.
- Swap the numbers over.
- Cancel down to the smallest possible ratio.

Examples:

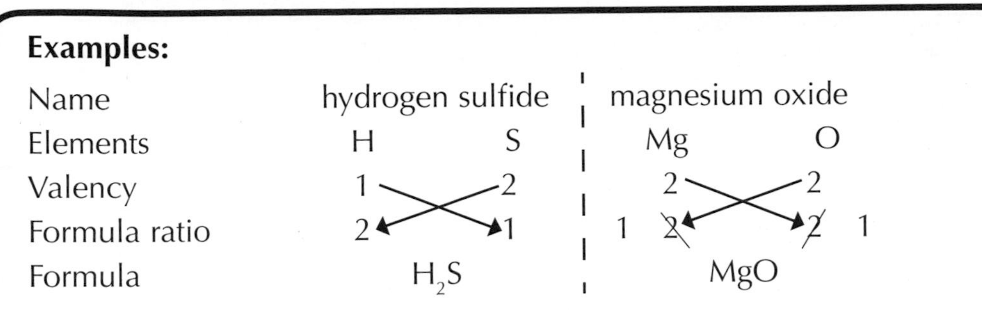

Name	hydrogen sulfide		magnesium oxide	
Elements	H	S	Mg	O
Valency	1	2	2	2
Formula ratio	2	1	~~2~~ 1	~~2~~ 1
Formula	H_2S		MgO	

The formulae of some compounds can be worked out from their names because they have a prefix to indicate the number of atoms: mono- = one; di- = two; tri- = three; tetra- = four; penta- = five, etc. Examples: carbon **mono**xide = **CO**; silicon **di**oxide = Si$\mathbf{O_2}$.

What does a formula tell us?

Hydrogen sulfide and carbon monoxide are covalent compounds consisting of individual molecules, and their formulae tell us exactly how many atoms of each element are in each molecule.

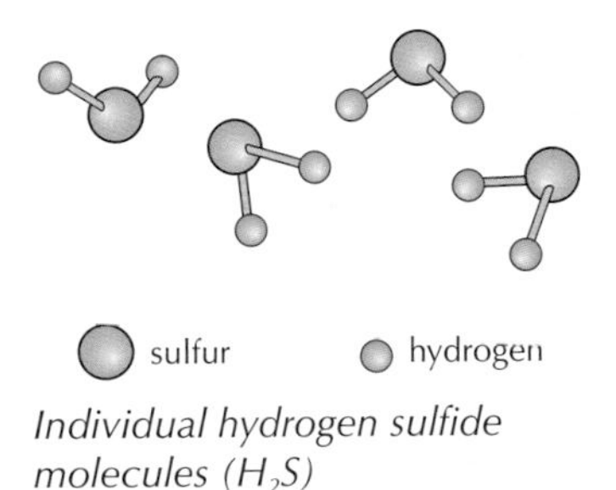

Individual hydrogen sulfide molecules (H_2S)

Silicon dioxide is a giant covalent network – there are no molecules. Its formula tells us the ratio of the atoms of each element in the compound – SiO_2 has an Si:O ratio of 1:2 (see page 15). The ratio rule applies to all covalent network compounds.

Magnesium oxide is ionic and exists as a giant crystal lattice. Its formula tells us the ratio of the ions in the compound – MgO has an Mg:O ratio of 1:1. The ratio rule applies to all ionic compounds.

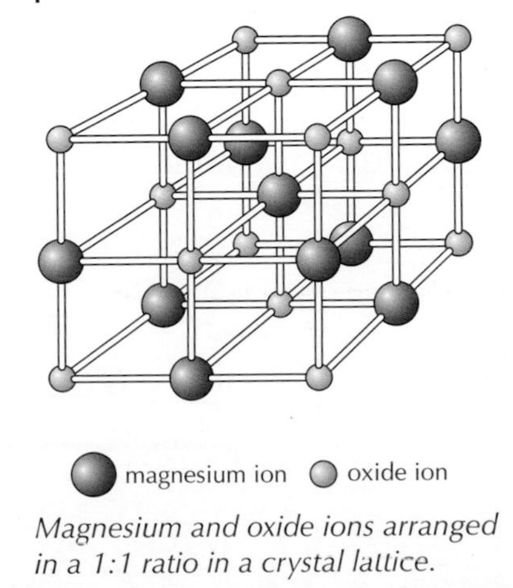

Magnesium and oxide ions arranged in a 1:1 ratio in a crystal lattice.

Formulae of compounds containing group ions

Some ions consist of more than one atom and are known as **group ions**, e.g. nitrate (NO_3^-), sulfate (SO_4^{2-}) and ammonium (NH_4^+).

The valency method can be used to work out chemical formulae for compounds containing group ions. The valency is the **number of charges** on the ion. For example the sulfate ion (SO_4^{2-}) has a 2^- charge, so its valency is 2.

TOP TIP

You don't have to memorise the group ions – they are listed in the SQA data booklet. You do need to know how to use them to work out a formula.

Examples:

Name	potassium sulfate		calcium nitrate	
Elements/group ion	K	SO_4^{2-}	Ca	NO_3^-
Valency	1	2	2	1
Formula ratio	2	1	1	2
Formula	K_2SO_4		$Ca(NO_3)_2$	

A **bracket** is always used when there is more than one group ion in a formula.

Roman numerals are used in chemical formulae to indicate the valency for elements that can have more than one valency – this is the case with many transition metals.

Examples:

Name	iron(III) carbonate		nickel(II) nitrate	
Elements/group ion	Fe	CO_3^{2-}	Ni	NO_3^-
Valency	3	2	2	1
Formula ratio	2	3	1	2
Formula	$Fe_2(CO_3)_3$		$Ni(NO_3)_2$	

TOP TIP

It is useful to learn Roman numerals up to 5: I = 1; II = 2; III = 3; IV = 4; V = 5.

When the charges are shown in the formula it is called an **ionic formula**.

Examples:

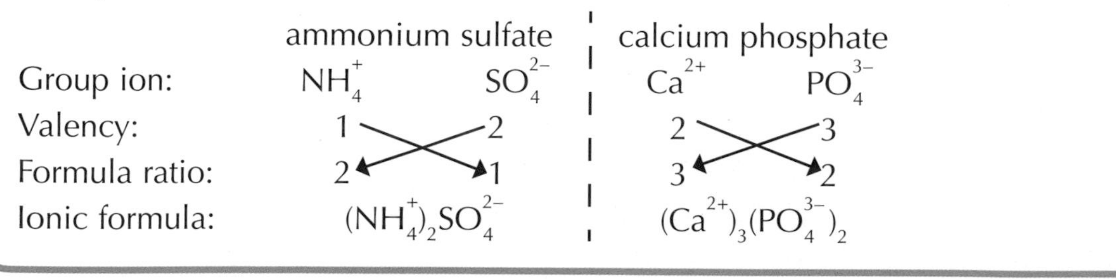

	ammonium sulfate		calcium phosphate	
Group ion:	NH_4^+	SO_4^{2-}	Ca^{2+}	PO_4^{3-}
Valency:	1	2	2	3
Formula ratio:	2	1	3	2
Ionic formula:	$(NH_4^+)_2SO_4^{2-}$		$(Ca^{2+})_3(PO_4^{3-})_2$	

A bracket is used in an ionic formula whenever there is more than one ion.

Formulae for ionic compounds can also be worked out by **balancing charges** – the overall charge on an ionic compound is neutral. For example, the charge on the ammonium ion (NH_4^+) is 1+ so when combined with the sulfate ion (SO_4^{2-}) there must be **two** NH_4^+ ions for every **one** SO_4^{2-} ion, i.e. $(NH_4^+)_2\ SO_4^{2-}$. Check calcium phosphate on page 20 – overall 6+ charges are balanced by 6– charges.

TOP TIP

Be careful with name endings in ionic compounds – they can look similar but mean different things:

- Potassium sulf**ide**: the **–ide** tells us there are two ions, one is sodium so the other must be the sulf**ide** ion – $\mathbf{S^{2-}}$.
  Ionic formula: $\mathbf{(K^+)_2S^{2-}}$. Formula without ion charges: $\mathbf{K_2S}$.
- Potassium sulf**ate**: the group ion table shows that sulf**ate** is $\mathbf{SO_4^{2-}}$.
  Ionic formula: $\mathbf{(K^+)_2SO_4^{2-}}$. Formula without ion charges: $\mathbf{K_2SO_4}$.
- Potassium sulf**ite**: the group ion table shows that sulf**ite** is $\mathbf{SO_3^{2-}}$.
  Ionic formula: $\mathbf{(K^+)_2SO_3^{2-}}$. Formula without ion charges: $\mathbf{K_2SO_3}$.

Quick Test

Work out formulae for the following. Use the data in the SQA data booklet to help you – it can be downloaded from the SQA website.

1. (a) barium, (b) fluorine
   (c) boron hydride, (d) barium oxide
   (e) magnesium sulfide, (f) calcium nitride
   (g) phosphorus pentachloride, (h) nitrogen dioxide.
2. (a) sodium carbonate, (b) magnesium sulfite,
   (c) potassium nitrate, (d) lithium hydroxide.
3. (a) copper(II) chloride, (b) silver(I) nitride,
   (c) iron(III) fluoride, (d) iron(II) sulfide.
4. (a) calcium hydroxide, (b) iron(III) sulfite,
   (c) magnesium phosphate, (d) ammonium carbonate.
5. Write ionic formulae for 4(a)–(d) above.

TOP TIP

Always show your working when working out a formula or doing a calculation. It lets the examiner clearly see how you got your answer.

Balancing chemical equations

Chemical equations

TOP TIP

Make sure you use the formula rules to work out formulae for elements and compounds in equations.

A chemical equation is a shorthand way of showing chemicals reacting and the new chemicals produced:

reactants → products

Chemical equations can be written in words and chemical formulae.

Example 1: Hydrogen reacts with chlorine to form hydrogen chloride.

Word equation: hydrogen (diatomic element) + chlorine (diatomic element) → hydrogen chloride (compound)

Formula equation: $H_2 + Cl_2 \rightarrow HCl$

Example 2: Copper reacts with silver(I) nitrate solution to produce silver and copper(II) nitrate solution.

Word equation: copper + silver(I) nitrate → silver + copper(II) nitrate

Formula equation: $Cu + AgNO_3 \rightarrow Ag + Cu(NO_3)_2$

Balanced chemical equations

TOP TIP

State symbols can be used to indicate the state in which the reactants and products exist: solid = (s), liquid = (ℓ), gas = (g), and solution = (aq).

In a **balanced equation** the total number of atoms of each element on the left-hand side of the equation equals the total on the right.

The equations in examples 1 and 2 are unbalanced because the number of atoms on each side of the equation is not equal.

In example 1: $H_2 + Cl_2 \rightarrow HCl$

This shows there are more hydrogen and chlorine atoms on the left-hand side than on the right. Atoms can't just disappear. The 'missing' hydrogen and chlorine atoms must have formed another hydrogen chloride molecule, as this is the only product.

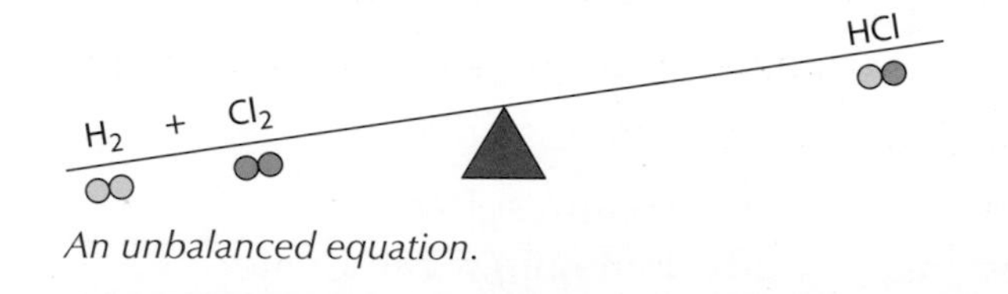

An unbalanced equation.

Balancing the equation:

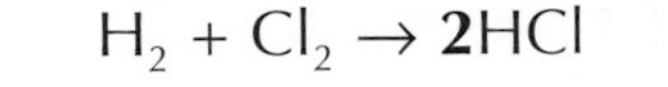

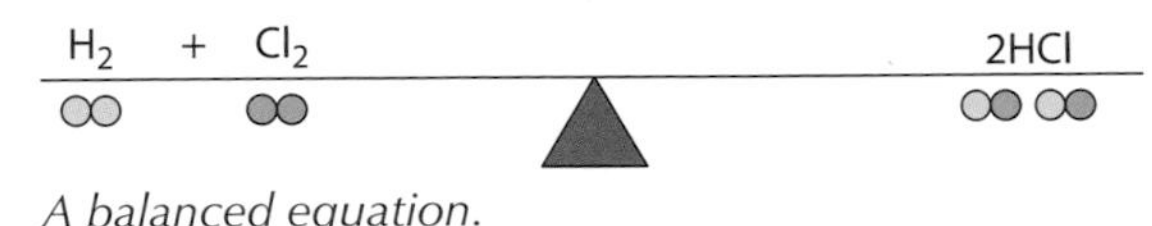

A balanced equation.

Formulae are never changed in order to balance an equation.

There are now the same number of atoms on each side of the equation.

When balancing an equation, numbers are put in front of formulae, never in between the symbols in the formula.

In example 2:

Unbalanced equation: $Cu + AgNO_3 \rightarrow Ag + Cu(NO_3)_2$

The unbalanced formula equation shows that there are two NO_3 groups on the right-hand side but only one on the left-hand side. To balance the two NO_3 groups on the right-hand side a 2 has to be put in front of the $AgNO_3$ on the left-hand side.

$$Cu + \mathbf{2}AgNO_3 \rightarrow Ag + Cu(NO_3)_2$$

The equation is still not balanced because there are now two Ag on the left-hand side. This is balanced by adding a 2 in front of the Ag on the right-hand side.

Balanced equation: $Cu + \mathbf{2}AgNO_3 \rightarrow \mathbf{2}Ag + Cu(NO_3)_2$

Quick Test

1. Write formulae equations for the following reactions:
   (a) barium chloride + magnesium sulfate → barium sulfate + magnesium chloride
   (b) iron + copper(II) nitrate → iron(II) nitrate + copper
   (c) copper(II) sulfate + sodium hydroxide → copper(II) hydroxide + sodium sulfate
2. Balance the following equations:
   (a) $Na + S \rightarrow Na_2S$
   (b) $K + O_2 \rightarrow K_2O$
   (c) $AgNO_3 + MgCl_2 \rightarrow AgCl + Mg(NO_3)_2$

Gram formula mass and the mole

Remember from National 4!

The relative formula mass of a **compound**, often referred to as formula mass, can be calculated from its formula by adding together the relative atomic masses (RAM) of all the atoms shown in the formula.

- For potassium sulfate: $\mathbf{K_2}$ **S** $\mathbf{O_4}$

$$(39 \times 2) + 32 + (4 \times 16) = 174$$

The formula mass of K_2SO_4 is **174**.

- For calcium nitrate: **Ca** **(N** $\mathbf{O_3)_2}$

$$40 + ((14 + (3 \times 16)) \times 2) = 164$$

The formula mass of $Ca(NO_3)_2$ is **164**.

TOP TIP

A table of RAM values for some common elements can be found in the SQA data booklet.

TOP TIP

Setting out your working this way means you are less likely to make a mistake and clearly shows how you have worked out the answer.

The mole

In order to do chemical calculations chemists use a quantity called the **mole,** often shortened to **mol**.

One mole of any substance is its formula mass in grams, i.e. the **gram formula mass (gfm)**.

So, using the examples above:

1 mol of K_2SO_4 has a gram formula mass of **174 g**, i.e. **1 mol = 174 g**.

1 mol of $Ca(NO_3)_2$ has a gram formula mass of **164 g**, i.e. **1 mol = 164 g**.

It is possible to have more than one mole of a substance and also fractions of a mole.

For example, **2 mol** of K_2SO_4 = 2 x 174 = **348 g** and **0.5 mol** of K_2SO_4 = 0.5 × 174 = **87 g**.

Generally, **mol = mass / gfm.**

From this, **mass = mol × gfm.**

Example 1: Calculate the number of **moles** of $Ca(NO_3)_2$ in 196 g.

Worked answer: mol = mass / gfm

= 196 / 164

mol = 1.2 mol

TOP TIP

You may wish to use the triangle to help you to remember the relationship between moles, mass and gram formula mass (gfm).

mass
mol | gfm

Example 2: Calculate the number of **moles** of K_2SO_4 in 43.5 g

Worked answer: mol = mass / gfm

= 43.5 / 174

= 0.25 mol

Example 3: Calculate the **mass** of K_2SO_4 in 0.45 mol.

Worked answer: mass = mol × gfm

= 0.45 × 174

mass = 78.3 g

Example 4: Calculate the **mass** of 1.3 mol of $Ca(NO_3)_2$

Worked answer: mass = mol × gfm

= 1.3 × 164

mass = 213.2 g

Quick Test

Check your answer to each question before moving on to the next one.

1. Calculate the gram formula mass of the following:
   (a) calcium carbonate ($CaCO_3$)
   (b) iron(III) hydroxide ($Fe(OH)_3$)
   (c) copper(II) nitrate ($Cu(NO_3)_2$)
   (d) ammonium sulfate ($(NH_4)_2SO_4$)

Use your answers to 1 (a)–(d) to answer questions 2 and 3.

2. Calculate the number of moles in each of the following:
   (a) 7.5 g of $CaCO_3$
   (b) 162 g of $Fe(OH)_3$
   (c) 18.75 g of $Cu(NO_3)_2$
   (d) 231 g of $(NH_4)_2SO_4$
3. Calculate the mass of each of the following:
   (a) 2.3 mol of $CaCO_3$
   (b) 0.2 mol of $Fe(OH)_3$
   (c) 1.4 mol of $(Cu(NO_3)_2$
   (d) 0.6 mol of $(NH_4)_2SO_4$

Connecting moles, volume and concentration in solutions

Mass and moles are connected – moles can be worked out from mass and gfm, and mass from moles and gfm.

When a substance is dissolved in water, a solution is formed. The **concentration** of the dissolved substance in the solution can be worked out if the number of moles (or mass) of the dissolved substance and the volume of solution formed are known.

concentration of solution = moles of substance dissolved / volume of solution made (in **litres**)

c = mol / v

From this, **mol** = **c × v** and **v** = **mol / c**

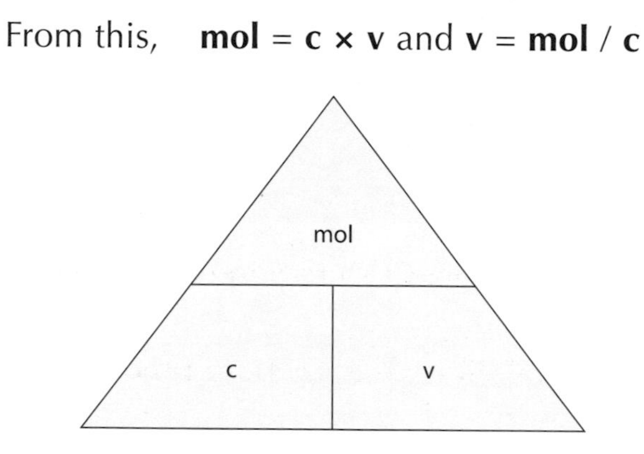

You may wish to use the concentration triangle to help you remember the connection between concentration, moles and volume.

Example 1: Calculate the **concentration** of the solution formed when 0.3 mol of sodium hydroxide is dissolved in water and the volume made up to 250 cm^3.

Worked answer: c = mol / v

= 0.3 / 0.25

c = 1.2 mol l^{-1}

Example 2: Calculate the number of **moles** in 50 cm^3 of a 0.1 mol l^{-1} solution of silver nitrate.

Worked answer: mol = c × v

= 0.1 × 0.05

mol = 0.005 mol

TOP TIP

Volume must be in litres. The unit of concentration is mol l^{-1} (moles per litre).

Example 3: Calculate the **volume** of a 0.15 mol l^{-1} solution that contains 0.45 mol of sodium chloride.

Worked answer: v = mol / c

= 0.45 / 0.15

v = 3.0 l

Example 4: Calculate the **concentration** of a solution, in **mol l^{-1}**, of potassium hydroxide (KOH) made when 1.7 g is dissolved in water and made up to 100 cm^3.

Worked answer: The answer has to be in **mol l^{-1}** so the mass in grams must first be converted to moles:

mol = m / gfm

= 1.7 / 56

= 0.03 mol

gfm of K O H: 39 + 16 + 1 = 56 g

Then, c = mol / v

= 0.03 / 0.1

c = 0.003 mol l^{-1}

Example 5: Calculate the **mass** of sodium carbonate (Na_2CO_3) required to prepare 100 cm^3 of a solution of concentration 0.25 mol l^{-1}.

Worked answer: In order to calculate mass in a solution, the number of moles must first be calculated.

mol = c × v

= 0.25 × 0.1

mol = 0.025 mol

Then, mass = mol × gfm

= 0.025 × 106

mass = 2.65 g

gfm of Na_2 C O_3: (2 × 23) + 12 + (3 × 16) = 106 g

Quick Test

1. Calculate the concentration, in mol l^{-1}, of the solution formed when 0.3 mol of lithium chloride is dissolved in a little water and made up to 500 cm^3 with water.
2. Calculate the concentration, in mol l^{-1}, of the solution formed when 3.5 g of potassium sulfate (gfm = 174 g) is dissolved in a little water and made up to 250 cm^3 with water.
3. Calculate the number of moles in 360 cm^3 of a 0.25 mol l^{-1} calcium nitrate solution.
4. Calculate the mass of ammonium nitrate (gfm = 80 g) required to prepare 250 cm^3 of a solution of concentration 0.7 mol l^{-1}.

Acids and bases

Remember from National 4!

pH is used as a way of indicating whether a solution is acid, alkali or neutral:

Acids: pH less than 7 Alkalis: pH greater than 7 Neutral: pH = 7

TOP TIP

Always pH, **never** Ph, PH or ph.

pH and hydrogen ion concentration

pH is a measure of the hydrogen ion concentration in solution.

In water and neutral solutions a very small proportion of the water molecules break down to form equal concentrations of hydrogen ions and hydroxide ions and the pH = 7.

This is known as the **dissociation of water** and is shown as:

$$[H_2O(\ell)] \rightleftharpoons [H^+(aq)] + [OH^-(aq)] \quad [\ ] = \text{concentration}$$

The $\rightleftharpoons$ symbol shows that the reaction is reversible – as the molecules break down to form ions most rejoin to form water molecules again.

How do acids and alkalis form in solution?

When a **non-metal oxide** dissolves in water an acidic solution is formed. This means there is an increase in the $[H^+(aq)]$ and a decrease in the $[OH^-(aq)]$. The higher the $[H^+(aq)]$, the more acidic the solution is and the lower the pH number.

Sulfur dioxide, one of the causes of acid rain, dissolves in water to form sulfurous acid. Some of the sulfur dioxide molecules dissolve in water to give hydrogen ions and sulfite ions.

$$\text{sulfur dioxide} + \text{water} \rightarrow \text{sulfurous acid}$$

$$SO_2(g) + H_2O(\ell) \rightarrow 2H^+(aq) + SO_3^{2-}(aq)$$

When a **soluble metal oxide** dissolves in water an alkaline solution is formed. This means there is an increase in the $[OH^-(aq)]$ and a decrease in the $[H^+(aq)]$. The higher the $[OH^-(aq)]$ the lower the $[H^+(aq)]$ and the higher the pH number.

Sodium oxide, for example, ionises completely in water to form sodium hydroxide.

$$(Na^+)_2\,O^{2-}(s) + H_2O(\ell) \rightarrow 2Na^+(aq) + 2OH^-(aq)$$

This diagram summarises the comparison of ion concentration in acid and alkaline solutions with water.

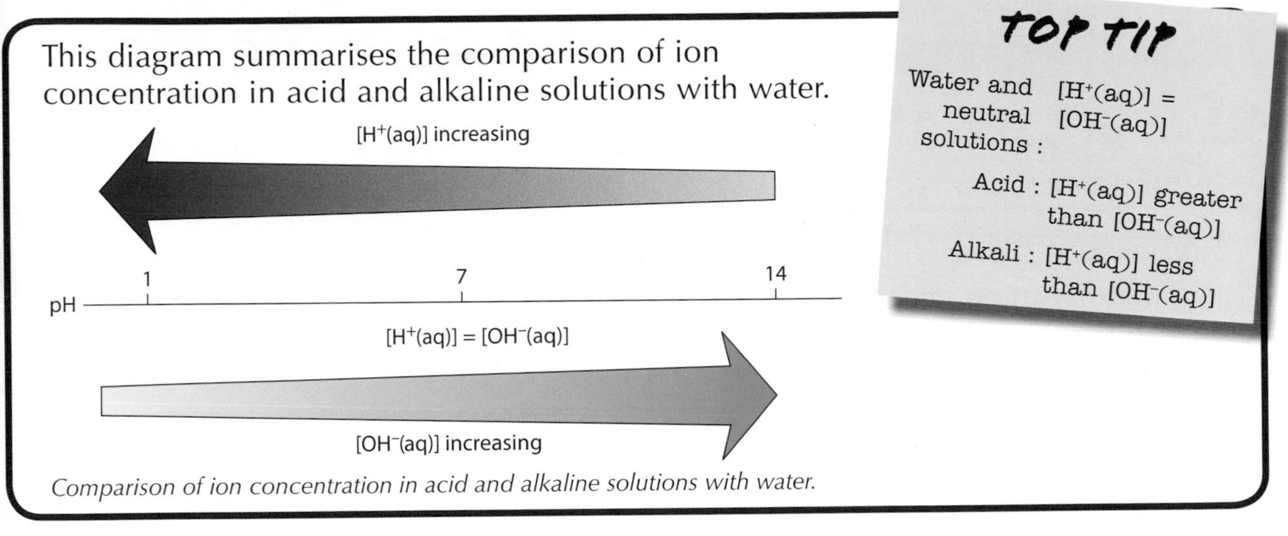

Comparison of ion concentration in acid and alkaline solutions with water.

TOP TIP

Water and neutral solutions : $[H^+(aq)] = [OH^-(aq)]$

Acid : $[H^+(aq)]$ greater than $[OH^-(aq)]$

Alkali : $[H^+(aq)]$ less than $[OH^-(aq)]$

Diluting acids and alkalis

When water is added to an acid the acid is diluted – its concentration decreases. This means the concentration of the $H^+(aq)$ ions decrease. The pH rises towards pH 7. The solution becomes less and less acidic.

During dilution of an alkali, because the concentration of the hydroxide ions is decreasing, the pH falls towards 7. The solution becomes less and less alkaline.

Quick Test

1. Complete the following summary. You may wish to use the word bank to help you.

In pure water and neutral solutions the concentration of (a) __________ ions is (b) __________ to the concentration of hydroxide (c) __________. When a (d) __________ oxide dissolves in water an acid solution is formed. The pH of an acid is (e) __________ 7 and the concentration of hydrogen ions is (f) __________ than the concentration of hydroxide ions. When a (g) __________ oxide dissolves in water an alkaline solution is formed. The pH of an alkaline solution is (h) __________ 7 and the concentration of hydrogen ions is (i) __________ than the concentration of hydroxide ions.

Word bank

above, below, equal, greater, hydrogen, ions, less, metal, non-metal

2. A solution has pH = 3.

(a) How will be the concentration of hydrogen ions compare with the hydroxide ions in the solution?

(b) The solution is diluted with water. Explain what will happen to the pH of the solution.

Neutralisation and volumetric titrations

Remember from National 4!

In a **neutralisation** reaction an acid reacts with a base to form water and a salt, and, if the base is a carbonate, carbon dioxide is also formed. Bases are metal oxides, hydroxides and carbonates.

Neutralisation

When an acid is neutralised the hydrogen ions are removed and replaced in solution by metal ions. Looking at the equations for neutralisation reactions helps us see exactly what is happening during the reactions.

Acid + metal hydroxide

hydrochloric acid + potassium hydroxide → water + potassium chloride

$$HCl(aq) + KOH(aq) \rightarrow H_2O(\ell) + KCl(aq)$$

$$H^+(aq) + Cl^-(aq) + K^+(aq) + OH^-(aq) \rightarrow H_2O(\ell) + K^+(aq) + Cl^-(aq)$$

Looking closely at the equation it can be seen that Cl^- and K^+ appear on both sides. This means that they have not taken part in the reaction. Ions that do not take part in a reaction are known as **spectator ions**. The ions that have taken part in the reaction are the H^+ and OH^- ions. Water is the only new product.

Rewriting the equation without the spectator ions:

$$H^+(aq) + OH^-(aq) \rightarrow H_2O(\ell)$$

The equation clearly shows the $H^+(aq)$ ions being removed from solution as $H_2O(\ell)$.

Acid + metal carbonate

sulfuric acid + lithium carbonate → water + lithium sulfate + carbon dioxide

$$H_2SO_4(aq) + Li_2CO_3(aq) \rightarrow H_2O(\ell) + Li_2SO_4(aq) + CO_2(g)$$

$$2H^+(aq) + SO_4^{2-}(aq) + 2Li^+(aq) + CO_3^{2-}(aq) \rightarrow H_2O(\ell) + 2Li^+(aq) + SO_4^{2-}(aq) + CO_2(g)$$

Spectator ions: SO_4^{2-} + Li^+. Reacting ions: $2H^+$ and CO_3^{2-}.

Rewriting the equation without the spectator ions:

$$2H^+(aq) + CO_3^{2-}(aq) \rightarrow H_2O(\ell) + CO_2(g)$$

Again, the equation clearly shows the $H^+(aq)$ ions being removed from solution as $H_2O(\ell)$.

Each of the acid reactions with a base clearly shows that in a neutralisation the only reaction taking place is the removal of the hydrogen ions as water.

TOP TIP

When an ionic compound is in solution the ionic lattice breaks down and the ions are free to move around. So, instead of writing the ionic formula for, say, potassium hydroxide as $K^+OH^-(aq)$, it is written as separate ions, $K^+(aq) + OH^-(aq)$.

Volumetric titrations

A neutralisation reaction between an acid and an alkali can be carried out accurately in an experiment called a **titration**.

- A titration involves using a **burette**, to gradually add accurately measured volumes of the acid into a conical flask containing alkali.
- The alkali is accurately measured using a **pipette**.
- A few drops of a chemical called an **indicator** are also added to the flask. Indicators change colour just as the neutralisation reaction is complete. This is the signal to stop the titration.
- The volume of acid added to neutralise the alkali is then read off the scale on the side of the burette.
- The volumes of acid and alkali can then be used to calculate the unknown concentration of an acid or alkali.

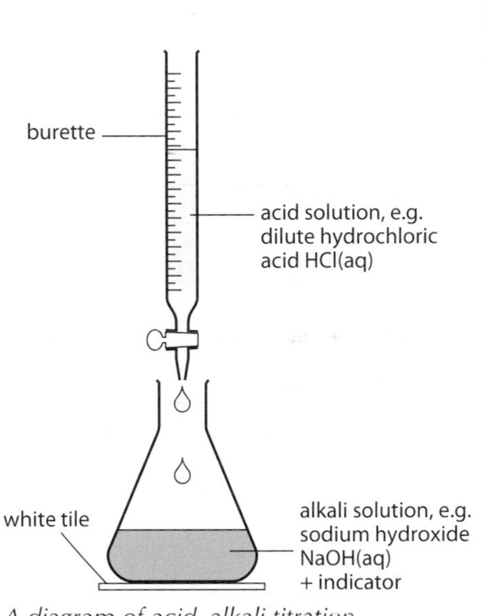

A diagram of acid–alkali titration.

Quick Test

1. The equation shows the neutralisation of potassium hydroxide with sulfuric acid.

   $H_2SO_4(aq) \; + \; 2KOH(aq) \; \rightarrow \; 2H_2O(\ell) \; + \; K_2SO_4(aq)$

   (a) Rewrite the equation showing the charge on each ion.

   (b) Rewrite the ionic equation without the spectator ions to show what happens to the hydrogen ions during neutralisation.

2. Complete the following summary. You may wish to use the word bank to help you.

   To carry out a titration of an acid with an alkali, the acid is poured into a (a)__________. The alkali is drawn up into a (b)__________ and emptied into a conical flask. A few drops of (c)__________ are added to the flask. The acid is carefully added to the flask and at the neutralisation point the indicator changes (d)__________. This is the signal to stop the (e)__________. The volume of acid needed to (f)__________ the alkali is then read off the scale on the side of the burette.

   Word bank

   burette, colour, indicator, neutralise, pipette, titration

Learning checklist

In this chapter you have learned

Rates of reaction

- the average rate of reaction = change in volume of gas / change in time or = loss in mass / change in time
- the data needed to calculate average rate can be obtained from graphs

Nuclide notation, isotopes and relative atomic mass (RAM)

- how to use nuclide notation for an atom and ion
- most elements have isotopes – not all atoms of an element have the same number of neutrons
- relative atomic mass (RAM) is the average mass of the isotopes and can be calculated using information from a mass spectrometer

Covalent bonding and shapes of molecules

- non-metal atoms are held together in a covalent bond because of the attraction of the nucleus of one atom for the outer electrons of another
- dot and cross diagrams can be used to show how atoms share a pair of electrons to form a covalent bond
- multiple bonds can be formed between the atoms in some covalent molecules
- the specific shapes of covalent molecules can be drawn

Structure and properties of covalent substances

- covalent molecular substances have low melting and boiling points because the forces of attraction between molecules are very weak, and not a lot of energy is needed to separate the molecules
- a covalent network is a giant three-dimensional structure in which the atoms are covalently bonded to each other
- covalent network substances have high melting and boiling points because the atoms are tightly held together by strong covalent bonds, and a lot of energy is needed to break the bonds

Structure and properties of ionic compounds

- ionic compounds exist as lattices in which electrostatic attractions hold the oppositely charged ions in a three-dimensional structure
- ionic compounds have high melting and boiling points because it takes a lot of energy to break the ionic bonds
- ionic compounds do not conduct electricity when solid because the ions cannot move but do when melted or in solution because the ions are free to move
- when ionic compounds dissolve in water the electrostatic attraction between ions is replaced by attractive forces between the ions and water molecules
- many ionic compounds are highly coloured compared to covalent substances

Chemical formulae and equations

- the formula of a covalent molecular substance tells us the exact number of atoms in each molecule
- the formula of a covalent network substance or an ionic compound tells us the ratio of the atoms or ions in the substance
- how to write chemical formulae for compounds containing group ions
- how to write ionic formulae
- how to write formulae equations for compounds containing group ions
- how to balance chemical equations

Gram formula mass and the mole

- the gram formula mass of a substance is its formula mass measured in grams
- the gram formula mass of a substance is known as the mole
- how to carry out calculations involving mass into moles and moles into mass

  mass = mol × gram formula mass

  mol = mass / gram formula mass

Connecting moles, volume and concentration in solutions

- the connection between mass, volume of solutions, concentration and moles

  c = mol / v; mol = c × v; v = mol / c

Acids and bases

- pure water contains H_2O molecules and equal concentrations of $H^+(aq)$ and $OH^-(aq)$ ions
- all aqueous solutions contain $H^+(aq)$ and $OH^-(aq)$ ions
- pH is a measure of the $H^+(aq)$ ion concentration in a solution
- the concentration of $H^+(aq)$ ions in acids is higher than in water
- the concentration of $H^+(aq)$ ions in alkalis is lower than in water
- the concentration of $OH^-(aq)$ ions in alkalis is higher than in water
- the concentration of $OH^-(aq)$ ions in acids is lower than in water
- diluting acids reduces the concentration of $H^+(aq)$ so the pH of the acid increases towards 7
- diluting alkalis reduces the concentration of OH^- (aq) so the pH of the alkali decreases towards 7

Neutralisation and volumetric titrations

- only the $H^+(aq)$ ions and $OH^-(aq)$ ions react in a neutralisation reaction
- a neutralisation reaction can be followed by volumetric titration
- coloured indicators are used in an acid / alkali titration to show when neutralisation has occurred

Alkanes

Remember from National 4!

Alkanes are saturated hydrocarbons with names ending in -ane. The names of straight chain alkanes can be worked out from their molecular and structural formulae.

Alkane structures

A family of compounds that have similar chemical properties and show a trend in physical properties, e.g. boiling point, and can be represented by the same general formula, is called a **homologous series**. Alkanes are a homologous series of hydrocarbons with the **general formula C_nH_{2n+2}** where **n = 1, 2, 3**, etc.

Butane has the molecular formula C_4H_{10}. Four carbons and ten hydrogens can be arranged to give two different structures.

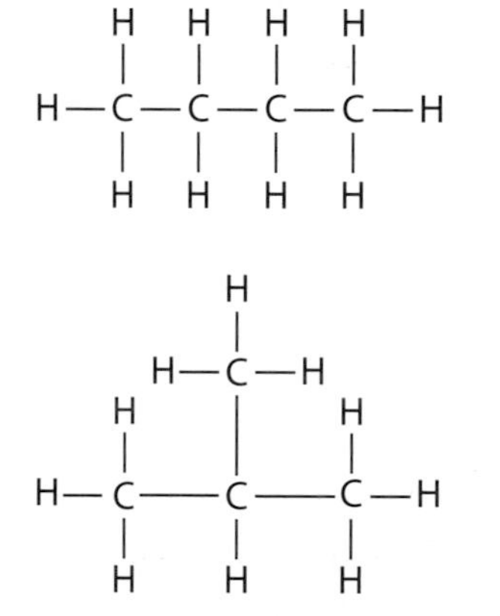

Structure 1: This arrangement is a **straight-chain** alkane.

Structure 2: This arrangement gives a **branched-chain** alkane.

The molecules above are described as **isomers**. Their chemistry is very similar but they have different physical properties. Alkanes with more than three carbon atoms have isomers.

Shortened structural formulae can be written for the structures above.

Structure 1: $CH_3CH_2CH_2CH_3$

Structure 2: $CH_3CH(CH_3)CH_3$

Branches are shown in brackets following the carbon they are attached to.

TOP TIP

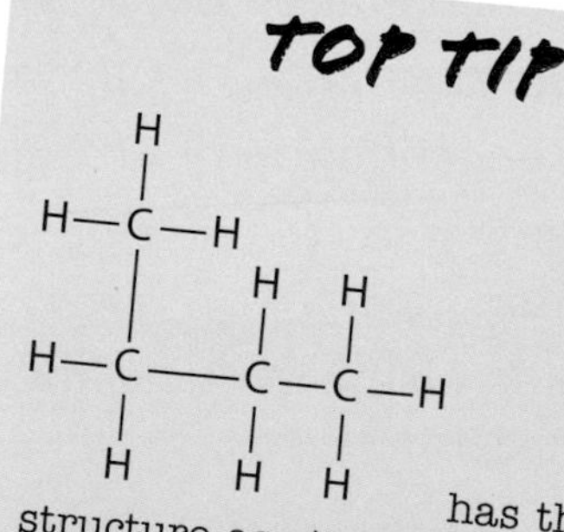

has the same structure as structure 1 so is not an isomer of 1. Although the way it is drawn makes it looks like there is a branch, there is not – what looks like a branch on the left-hand side is part of the straight chain.

Physical properties

Although the isomers of alkanes have similar chemical properties, their physical properties are different. Branching in liquid hydrocarbons causes the intermolecular forces to decrease. This means less energy is needed to separate branched molecules. This results in branched-chain isomers having **lower boiling points** and **higher volatility** than the straight-chain hydrocarbon – the more volatile a molecule is, the easier it is for it to form a gas.

Isomer	Pentane	2-Methylbutane	2,2-Dimethylpropane
Structure	$CH_3CH_2CH_2CH_2CH_3$ (displayed formula)	$CH_3CH(CH_3)CH_2CH_3$ (displayed formula)	$C(CH_3)_4$ (displayed formula)
Boling point (°C)	36	27	11

Boiling points of the isomers of pentane

Using branched alkanes

Branched-chain alkanes are particularly important in petrol manufacture where they are used to improve how smoothly the fuel burns. Petrol is a mixture of hydrocarbons in the C_5 to C_{12} range that include branched-chain alkanes, which burn more smoothly in a petrol engine than straight-chain alkanes.

The volatility of alkanes used in petrol is also important. For petrol to burn in an engine it must vaporise and mix with air. The volatility of the petrol is critical. In winter, more volatile components are added to petrol so that it vaporises more easily.

Quick Test

1. The alkanes are a **homologous series,** which can be represented by the general formula C_nH_{2n+2}. Alkanes with more than three carbon atoms have **isomers**.
   (a) Give the meaning of the terms in bold type.
   (b) Work out the formula of the alkane with nine carbon atoms.
2. Explain why a branched-chain isomer of butane has a lower boiling point than the straight-chain isomer.

Naming and drawing branched alkanes

Naming from a structural formula

TOP TIP

Make sure you know the names of the first eight alkanes, as this will help you name the branched-chained alkanes.

Branched alkanes are named **systematically**.

1. Identify and name the longest hydrocarbon chain.
2. Identify the branches (side chains).
   For example: -CH_3 is **methyl**; -C_2H_5 is **ethyl**
3. Prefixes are used if there is more than one side chain of the same type, e.g. **di**- is used if there are two of the same type, **tri**- if there are three, etc.
4. To indicate the position of the branches on the main chain, number the carbon atoms from the end of the main chain nearer a branch.

Using these rules, the systematic name of the branched chain isomer of butane (see structure 2 on page 34) is:

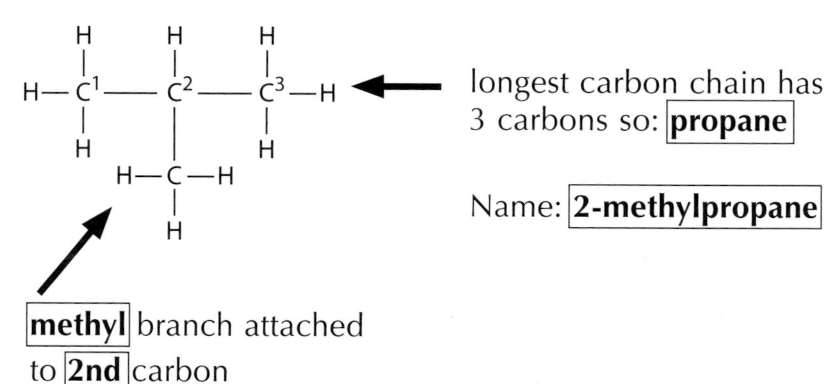

Pentane, C_5H_{10}, has **three isomers**, two of which are branched. The structures and sytematic names of the branched isomers are detailed below, the first one showing how the above rules are applied:

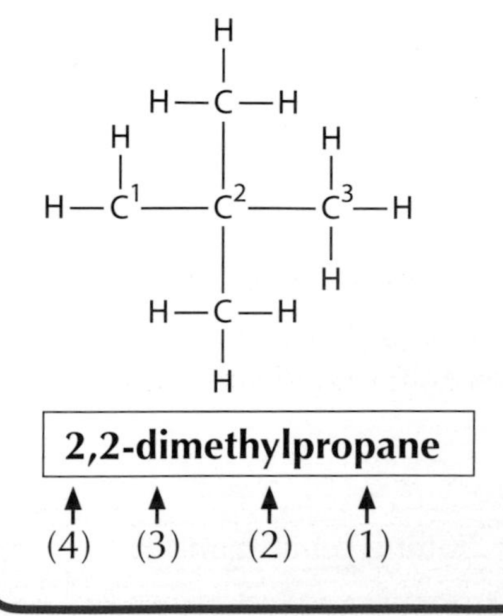

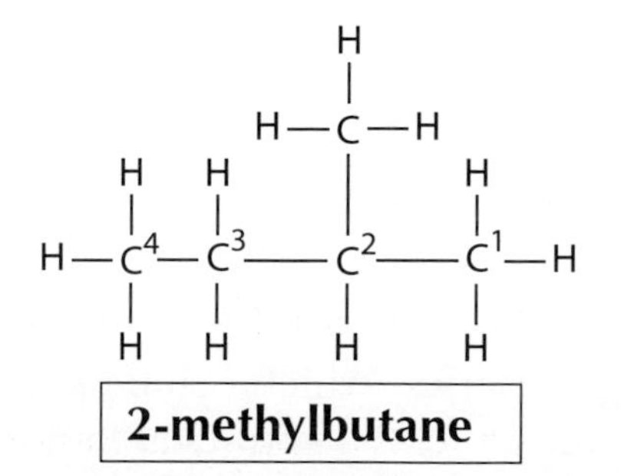

Note C atoms numbered from right-hand side to give branch lowest number

Drawing an isomer from its name

1. Identify the longest straight chain.
2. Identify the branches and how many.
3. Identify the position of the branches.

Worked example:

Draw the structure for **2,2,4-trimethylpentane**.

1. The longest straight chain is pentane – it has five carbon atoms.

   C—C—C—C—C

2. There are three methyl branches: 3 × CH_3.
3. Two methyl groups are attached to the second carbon in the chain and one to the fourth carbon.

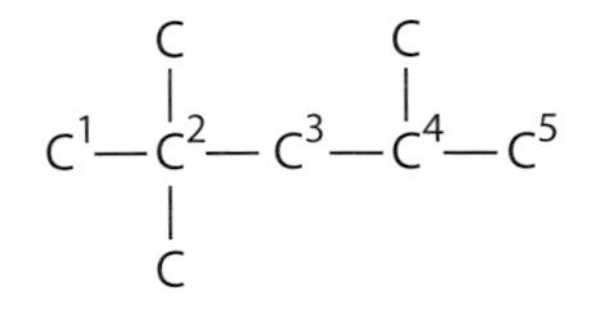

Structural formula

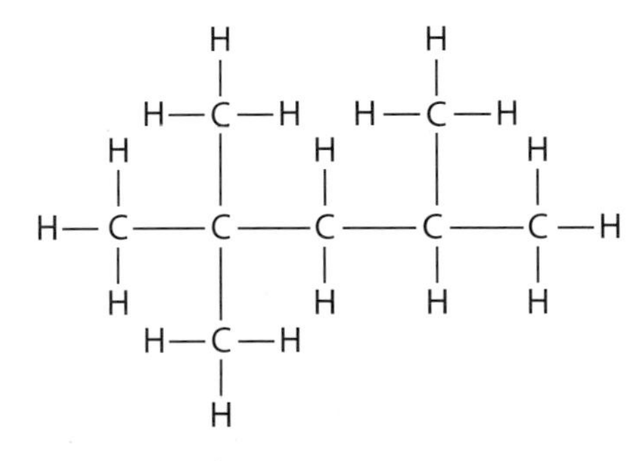

Shortened structural formula: $CH_3C(CH_3)_2CH_2CH(CH_3)CH_3$

Molecular formula: C_8H_{18}.

2,2,4-trimethylpentane is an **isomer** of octane.

Quick Test

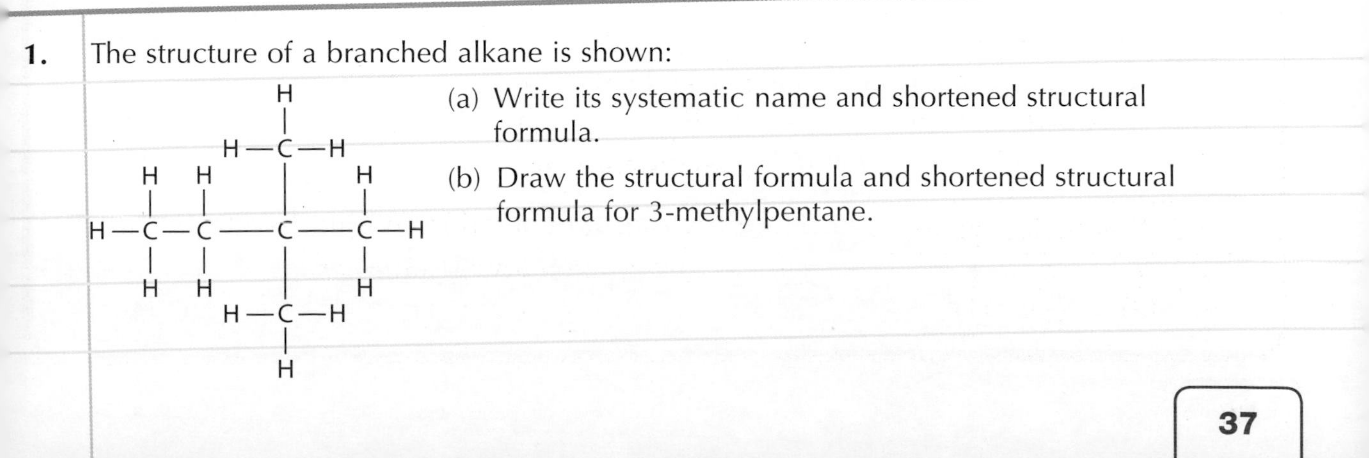

1. The structure of a branched alkane is shown:

   (a) Write its systematic name and shortened structural formula.

   (b) Draw the structural formula and shortened structural formula for 3-methylpentane.

Cycloalkanes

As the name suggests, **cycloalkanes** are hydrocarbon compounds with the carbons arranged in a **ring**. Like the alkanes these molecules are **saturated**, i.e. the carbons are joined to each other by single carbon-to-carbon bonds. The cycloalkanes are named simply by putting the prefix cyclo- in front of the name of the straight-chain alkane with the same number of carbon atoms.

Name	Molecular formula	Structural formula	Shortened structural formula
Cyclopropane	C_3H_6	H H C H—C—C—H H H	CH_2 H_2C—CH_2
Cyclobutane	C_4H_8	H H H—C—C—H H—C—C—H H H	H_2C—CH_2 H_2C—CH_2
Cyclopentane	C_5H_{10}	H H H C H H—C C—H C—C H H H H	CH_2 H_2C CH_2 H_2C—CH_2
Cyclohexane	C_6H_{12}	H H H H C—C H H C C H H C—C H H H H	H_2C—CH_2 H_2C CH_2 H_2C—CH_2

The first four cycloalkanes

Counting carbons and hydrogens gives $\mathbf{C_nH_{2n}}$ as the **general formula** for cycloalkanes where **n = 3, 4, 5,** etc. Cycloalkanes with more than three carbons have **isomers** in the same homologous series, e.g. cyclobutane.

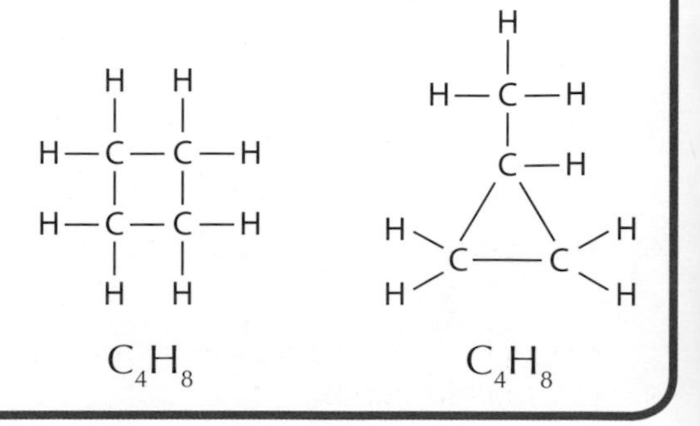

Cycloalkanes have the same general formula as alkenes and so are isomers, e.g. cyclopropane and propene:

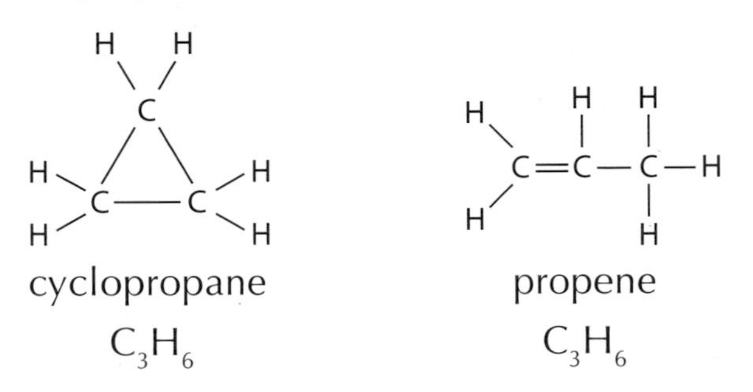

cyclopropane C_3H_6 — propene C_3H_6

Cycloalkanes are similar to alkanes in their general chemical and physical properties, but they have higher **boiling points** and **melting points** than alkanes. This is due to stronger **forces** of attraction between the molecules because the ring shape allows for a larger area of contact.

The alkane isomers of cyclobutane are shown on page 40.

Uses of cycloalkanes

Cycloalkanes are important constituents in petrol. Like branched alkanes they burn more smoothly than straight-chain alkanes.

Other uses of cycloalkanes are: solvents for paints and varnishes, feedstocks for making other chemicals, blowing agent for making polyurethane foam.

Quick Test

1. The general formula for the cycloalkanes is C_nH_{2n}.
(a) Work out the molecular formula for the cycloalkane with seven carbon atoms.
(b) Name the cycloalkane with seven carbon atoms.
(c) Draw the structural formula for the cycloalkane with seven carbon atoms.

2. (a) Draw the structural and shortened structural formulae for a cycloalkane isomer of cyclopentane.
(b) Draw the structural formula for a possible straight-chain alkene isomer of cyclopentane.

Alkenes

Remember from National 4!
Alkenes have a double C=C bond, and their name ends in –ene. General formula is C_nH_{2n} where n = 2, 3, 4 etc. Alkenes up to C_8 can be named. Alkenes are made by cracking alkanes.

TOP TIP

Knowing the names of the alkenes up to C_8 will help you with the **systematic naming** of the alkenes.

Naming isomers

Alkenes with more than three carbons in their molecules have **isomers** in the same homologous series. Not only do they have branches, but the position of the double bond can vary.

Example: Butene has two straight-chain isomers and one branched:

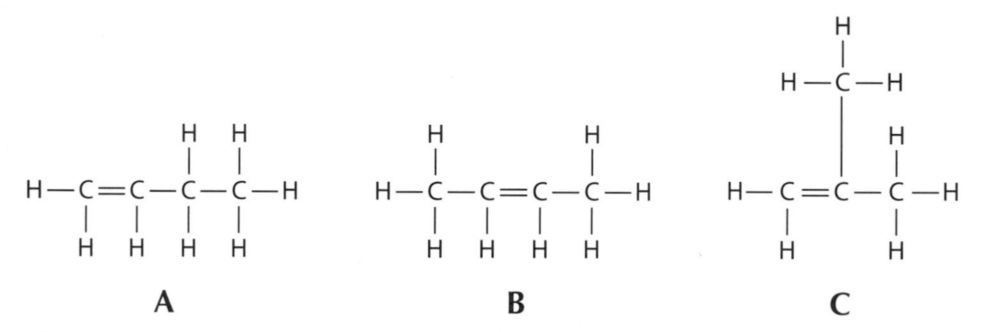

The structures are named in a similar way to alkanes, but the position of the double bond has to be included:

1. Identify the longest chain containing the double bond and name it.
2. Number the chain from the end nearer the double bond – the number of the first carbon atom with the double bond goes in the middle of the name.
3. Identify the branches (side chains) and how many there are. Prefixes are used if there is more than one side chain of the same type (di– is used if there are two of the same type, tri– if there are three, etc.).
4. Indicate the position of the branches on the main chain at the beginning of the name.

The isomers of butene, **A, B** and **C** (above), are named using the rules:

A: 1. longest chain: butene

2. double bond on 1st carbon: **but-1-ene**

B: 1. longest chain: butene

2. double bond on 2nd carbon: **but-2-ene**

C: 1. longest chain: propene

2. double bond on 1st carbon: prop-1-ene

3. one methyl branch: methylprop-1-ene
4. methyl on 2nd carbon: **2-methylprop-1-ene**

Structural formulae from names

Example: Draw the structural formula for **2,3-dimethylpent-1-ene.**

Worked answer:

1. Identify the longest straight chain with the double bond: pent– = 5 carbons.
2. Identify the position of the double bond: pent-1-ene = between 1st and 2nd carbon.
3. Identify branches and how many: dimethyl = two methyl branches.
4. Identify the position of the branches: 2,3- = on 2nd and 3rd carbons.

Structural formula:

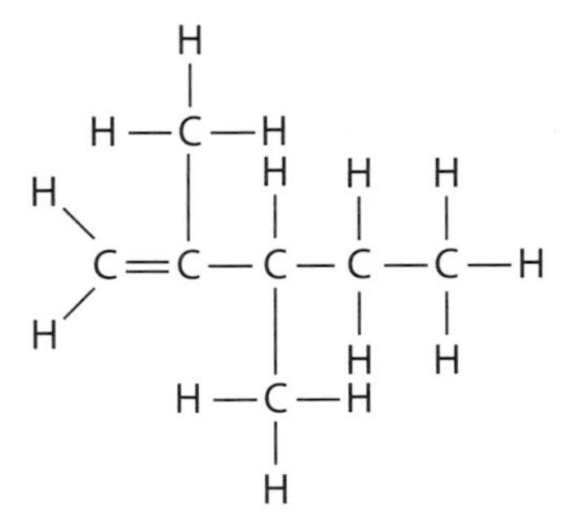

Shortened structural formula: $CH_2C(CH_3)CH(CH_3)CH_2CH_3$

Molecular formula: C_7H_{14}

Alkenes and cycloalkanes have the same general formula (C_nH_{2n}), so can be isomers. Cyclobutane (C_4H_8) would be a fourth isomer of butene (C_4H_8).

Quick Test

1. The structural formulae for three isomers of hexene (A, B and C) are shown:

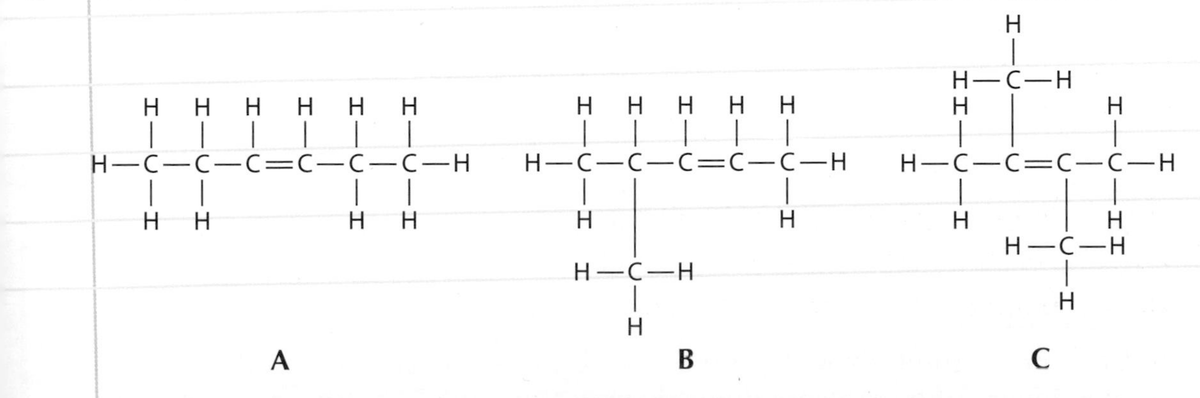

(a) Write systematic names and shortened structural formulae for A, B and C.

(b) Draw the structural formula for the isomer 3,3-dimethylbut-1-ene.

Reactions of alkenes

Remember from National 4!

Alkenes are unsaturated. They can be distinguished from alkanes by the way in which they react with bromine solution.

Addition reactions

Alkenes are **very reactive** due to the presence of the C=C double bond in the molecules. They are therefore important starting materials (feedstocks) in the chemical industry, used to make other important chemicals.

Reaction with halogens

Alkenes can be distinguished from alkanes using **bromine solution**.

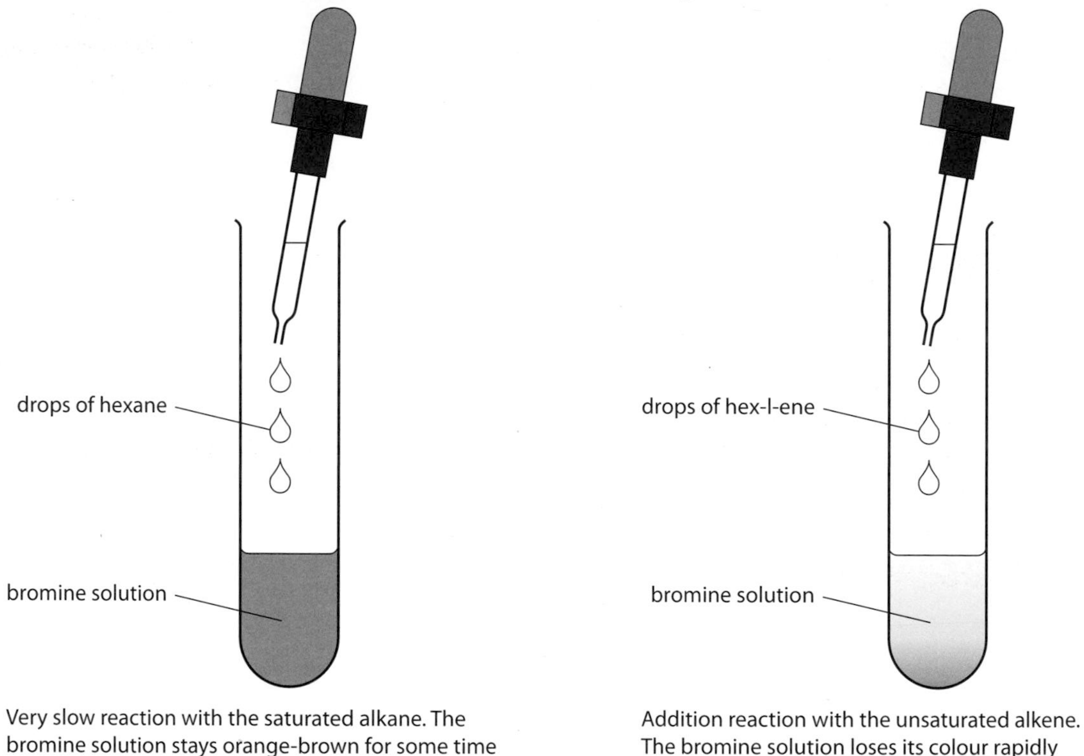

Comparing the reaction of hexane and hex-1-ene with bromine solution.

The reaction of hex-1-ene with bromine is an example of an **addition reaction** – the bromine atoms have added to the alkene. One of the bonds of the carbon-to-carbon double bond is broken and new atoms (or groups) join to the carbon chain. The product molecule is **saturated.** Other halogens add in the same way.

Adding hydrogen

Hydrogen can be added across the double carbon-to-carbon bond in an alkene (unsaturated) giving the corresponding alkane (saturated). The process is also known as **hydrogenation**.

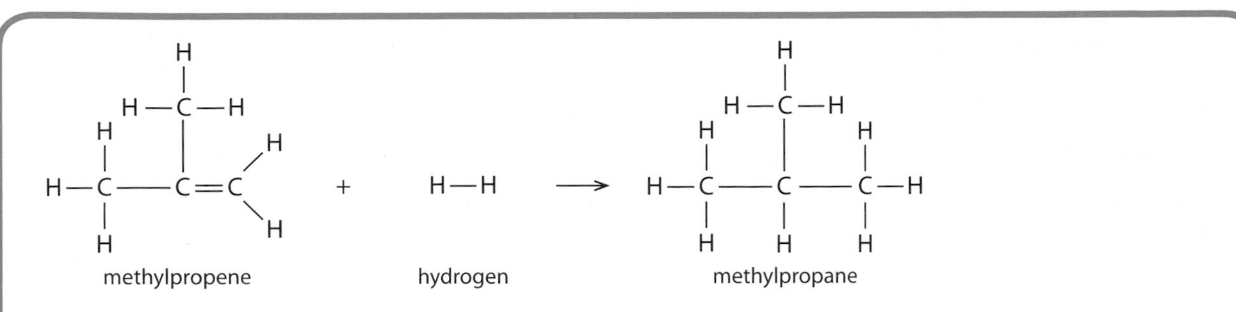

methylpropene hydrogen methylpropane

Adding water

The addition reaction, which involves adding water across a double carbon-to-carbon bond to form an alcohol, is also referred to as **hydration**.

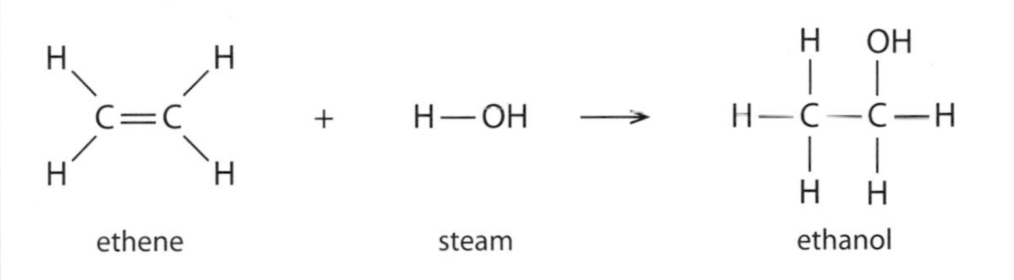

ethene steam ethanol

TOP TIP

When alkenes react, the reactants always add to the carbons with the double bond.

Making plastics

The most important use of alkenes is as feedstocks in the plastics industry. This is dealt with fully in Unit 3: Addition polymerisation.

Quick Test

1. Pent-1-ene reacts with chlorine to form 1,2-dichloropentane.

   $C_5H_{10} + Cl_2 \rightarrow C_5H_{10}Cl_2$

   (a) Draw structural formulae to show how the chlorine molecule reacts with pent-1-ene.

   (b) Name this type of reaction.

2. (a) But-1-ene reacts with hydrogen in an addition reaction.

   $C_4H_8 + H_2 \rightarrow X$

   (i) Give another name for this reaction.

   (ii) Write the molecular formula for compound X and name it.

   (iii) To which homologous series does X belong?

   (b) But-1-ene also reacts with steam in an addition reaction to produce compound Y.

```
   H  H  H  H             O
   |  |  |  |            / \
H—C=C—C—C—H    +    H     H   ——>   Y
         |  |
         H  H
```

   (i) Give another name for this reaction.

   (ii) Draw a possible structure for the compound Y.

Alcohols

Ethanol is the second member of a **homologous series** of **alcohols**, known as alkanols.

The names and formulae of the alkanols are related to the names and formulae of the corresponding alkanes. In the alcohols a hydrogen in the alkane molecule has been replaced by an oxygen bonded to hydrogen. This is known as a **hydroxyl group (–OH)**. All alcohols contain a hydroxyl group. This is the **functional group** for alcohols and is responsible for the similarities in properties of the members of the homologous series. The 'e' from the end of the alkane name is replaced with 'ol' to give the name of the alcohol.

TOP TIP

Don't mix up the names hydroxide and hydroxyl. Hydrox**ide** is the **OH^- ion**. Hydrox**yl** is the **–OH group**.

Alkane	Molecular formula	Alcohol	Molecular formula	Structural formula
Methane	CH_4	Methan**ol**	CH_3**OH**	H | H—C—OH | H
Ethane	C_2H_6	Ethan**ol**	C_2H_5**OH**	H H | | H—C—C—O—H | | H H
Propane	C_3H_8	Propan**ol**	C_3H_7**OH**	H H H | | | H—C—C—C—OH | | | H H H

Names and formulae for alcohols based on the corresponding alkanes

From the information in the table the general formula for the alcohols can be worked out. The general formula for the alcohols based on alkanes is $\mathbf{C_nH_{2n+1}OH}$.

Naming alcohol isomers

Alcohols with more than two carbon atoms have isomers – the position of the hydroxyl group can change. There are two isomers of propanol, C_3H_7OH.

TOP TIP

Alcohols have straight-chain and branched-chain isomers but you do not need to be able to draw or name branched structures.

Full structural formulae

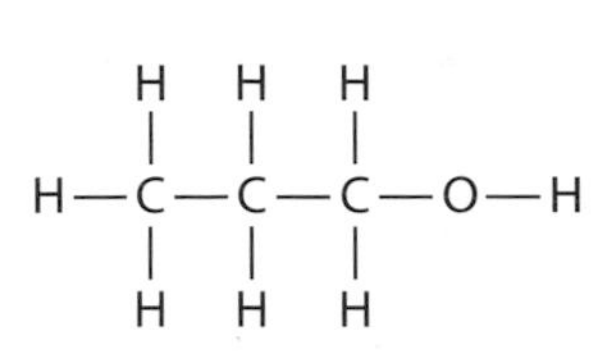

Structure 1

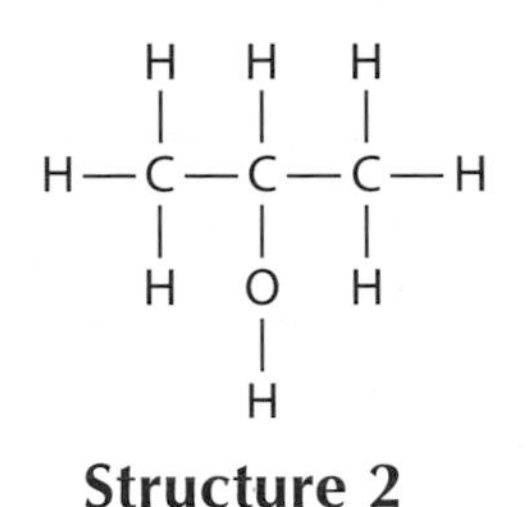

Structure 2

Shortened structural formulae

$CH_3CH_2CH_2OH$ $CH_3CH(OH)CH_3$

The name of an alcohol specifies where the hydroxyl group is attached to the carbon chain. The rules for naming are similar to those for naming alkenes (see Unit 2, page 40) but here the -OH group is given the lowest number.

> **TOP TIP**
>
> You will find it helpful when naming alcohols to know the names of the first eight alkanes.

Structures 1 and 2 can be named systematically.

Structure 1:

1. longest chain: propanol
2. hydroxyl on 1st carbon: **propan-1-ol**

Structure 2:

1. longest chain: propanol
2. hydroxyl on 2nd carbon: **propan-2-ol**

Formulae can be worked out from systematic names.

Worked example for **pentan-2-ol**.

1. The longest straight chain with the hydroxyl group = pentanol = 5 carbons.
2. The position of the hydroxyl group: 2 = 2nd carbon.

Structural formula:

```
      H
      |
   H  O  H  H  H
   |  |  |  |  |
H—C—C—C—C—C—H
   |  |  |  |  |
   H  H  H  H  H
```

Shortened structural formula: $CH_3CH(OH)CH_2CH_2CH_3$

Molecular formula: $C_5H_{11}OH$

Quick Test

1. Draw the structural formulae for the two straight chain isomers of butanol (C_4H_9OH) and name them.
2. Draw the structural formula for hexan-3-ol and work out its molecular formula.

Properties of alcohols

Properties

The alcohols are a homologous series, so show a gradual change in physical properties. The table shows the trend in melting and boiling points for straight-chain alcohols with the hydroxyl group attached to an end carbon of the chain.

Alcohol	Shortened structural formula	Melting point/°C	Boiling point/°C
Methanol	CH_3OH	−98	65
Ethanol	CH_3CH_2OH	−117	78
Propan-1-ol	$CH_3CH_2CH_2OH$	−127	97
Butan-1-ol	$CH_3(CH_2)_2CH_2OH$	−90	116
Pentan-1-ol	$CH_3(CH_2)_3CH_2OH$	−79	137

Melting points and boiling points of the first five alcohols

Explaining the trend in boiling points

The boiling points of the alcohols increase as the number of carbon atoms in the molecules increases. The longer the hydrocarbon chains, the greater the forces of attraction between molecules. This means more energy will be required to separate bigger molecules and so boiling points will increase.

There are other forces of attraction between alcohol molecules that are not present in alkanes of similar mass. This is due to the hydroxyl groups in alcohols. This results in alcohols having higher boiling points than alkanes of similar mass.

Propane: C_3H_8 (**formula mass: 44**)
bp = −44°C

Ethanol: C_2H_5OH (**formula mass: 46**)
bp = 78°C

Solubility in water

Small alcohol molecules like methanol and ethanol are very soluble and mix completely with water. For a molecule to dissolve in water it must be able to interact with water molecules. The presence of the –OH group on the alcohol results in strong forces of attraction forming between it and the –OH groups on the water molecules and so the small alcohol molecules mix easily with the water. The diagram shows how methanol molecules interact with water molecules:

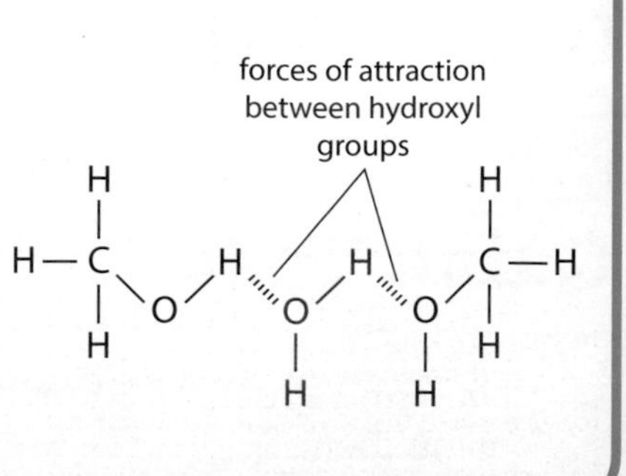

In **larger** alcohol molecules like octan-1-ol, the forces of attraction between the hydrocarbon parts of the alcohol molecules are stronger than the forces between the –OH groups, which makes it more difficult for large alcohol molecules to dissolve in water. The graph shows how the solubility of alcohols decreases as the number of carbon atoms in the molecule increases.

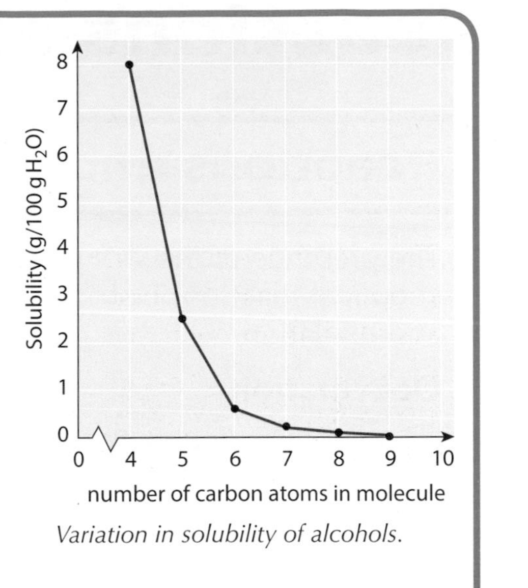

Variation in solubility of alcohols.

Extending the graph to include ethanol (C_2H_5OH) would show it to have a high solubility in water. Alcoholic drinks such as whisky contain ethanol diluted with water.

Bottle of whisky.

Quick Test

1. Look at the table of melting and boiling points of the alcohols.
   (a) Predict the boiling point of hexan-1-ol.
   (b) Explain the trend in the boiling points of the alcohols.
2. Explain why nonan-1-ol ($C_9H_{17}OH$) is insoluble in water but methanol is very soluble.

Reactions and uses of alcohols

Reactions and uses

The presence of the –OH group in alcohols means they undergo different types of chemical reaction. Two types of reaction of industrial importance are dehydration and oxidation.

Dehydration

Water can be removed from an alcohol to produce an alkene. The reaction is known as **dehydration**.

If an alcohol is heated with a dehydrating agent then water will be removed giving the corresponding alkene. Heating ethanol with a dehydrating agent gives ethene.

There is the possibility of using this reaction as an industrial source of ethene from ethanol produced from sugar cane, instead of cracking alkanes. Ethene is an important feedstock for the plastics industry.

Oxidation

Alcohols with the hydroxyl attached to an end carbon can be oxidised to give carboxylic acids. This happens naturally in wines when the ethanol reacts with oxygen from the air producing the acid, ethanoic acid. This sours the wine.

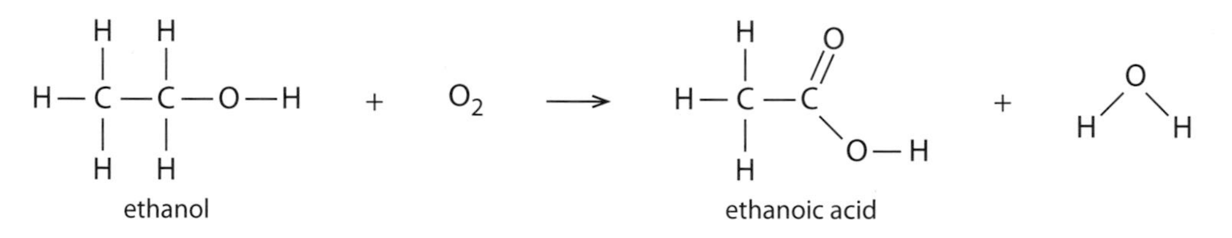

Fuels

Alcohols are extremely flammable and can be used as fuels. Methanol and ethanol burn cleanly to give carbon dioxide and water.

$$C_2H_5OH(\ell) + 3O_2(g) \rightarrow 2CO_2(g) + 3H_2O(\ell)$$
ethanol

Both methanol and ethanol can by used as fuels in car engines. Ethanol can be made by fermenting sugar cane, which is a renewable energy source.

Solvents

Ethanol is an effective solvent. After water it is the most important solvent used by industry.

Ethanol, as well as being used extensively in industry, is a common solvent in perfumes, food flavourings and many medical preparations.

Propan-2-ol dissolves a wide range of compounds and is used in cleaning fluids to dissolve oils. It is used to clean computer keyboards and monitor screens. It evaporates quickly and is relatively non-toxic, compared to some other solvents.

Propan-2-ol is also the alcohol used in many hand gels and disinfectant wipes.

Making alcohols

Alcohols can be made by the hydration of alkenes (see page 43)

Quick Test

1. Methanol and ethanol can be used as fuels.

(a) State two reasons for using ethanol as a fuel.

(b) Write a balanced equation for the combustion of methanol.

(c) State one other use for alcohols.

2. The flow diagram summarises chemical reactions involving propanol.

$$\text{propene} \underset{(h)}{\overset{(a)}{\rightleftarrows}} \text{proponol} \xrightarrow{(c)} \text{propanoic acid}$$

Name processes (a), (b) and (c).

Carboxylic acids and esters

Carboxylic acids are a homologous series that contain the **carboxyl group, –COOH**. The structure of the carboxyl group is:

```
     O
    //
 —C
    \
     O—H
```

The carboxyl group occurs at the end of a chain of carbons.

The general formula for the carboxylic acids based on alkanes can be written as $\mathbf{C_nH_{2n+1}COOH}$ where **n=0, 1, 2 etc.**

Using the general formula, the first carboxylic acid in the series (n=0) has the molecular formula **HCOOH**. The name is based on the corresponding alkane. The 'e' is dropped and replaced with '–oic acid': **methanoic acid**.

```
      O
     //
 H—C
     \
      O—H
```

methanoic acid

The second member of the series, when n=1, the molecular formula is: $\mathbf{CH_3COOH}$. The molecule has two carbon atoms so the name is **ethanoic acid**:

```
    H   O
    |  //
 H—C—C
    |   \
    H    O—H
```

ethanoic acid

TOP TIP

The carboxyl group has a carbon atom in it, which must be included when working out the name of a carboxylic acid.

Using carboxylic acids

A major use of **methanoic acid** is as a **preservative** and **antibacterial** agent in farm animal feed. Methanoic acid is also known as formic acid. The name comes from 'formica', the latin for ant. Ant and bee stings contain this acid.

Vinegar is probably the most well-known substance to contain a carboxylic acid. Vinegar is a solution of **ethanoic acid** (CH_3COOH) in water. It is widely used as a food preservative and flavouring.

Citric acid is a carboxylic acid found in citrus fruits such as lemon. **Lactic acid**, found in yoghurts, is made by fermenting lactose, a sugar found in milk. Lactic acid is also produced by the body. This causes the burn that people feel during strenuous exercise.

Vinegar is a solution of ethanoic acid.

Properties

Like the alcohols, straight-chain carboxylic acids show a gradual change in some physical properties.

Carboxylic acid	Shortened structural formula	Boiling point/°C	Solubility in water/g l^{-1}
Methanoic acid	HCOOH		
Ethanoic acid	CH_3COOH		
Propanoic acid	CH_3CH_2COOH		
Butanoic acid	$CH_3(CH_2)_2COOH$	Increasing	Decreasing
Pentanoic acid	$CH_3(CH_2)_3COOH$		
Hexanoic acid	$CH_3(CH_2)_4COOH$		
Heptanoic acid	$CH_3(CH_2)_5COOH$		
Octanoic acid	$CH_3(CH_2)_6COOH$		

Variation in boiling point and solubility of carboxylic acids

Boiling point: as the molecules get bigger, the forces of attraction between the molecules increase. This means the amount of energy needed to separate the molecules increases so the boiling point increases.

Solubility: small carboxylic acid molecules such as methanoic acid are very soluble and mix completely with water. The small molecules are able to interact with water molecules.

In **larger** carboxylic acid molecules, such as octanoic acid, the long hydrocarbon part of the molecules makes it more difficult for them to dissolve in water.

As the name suggests the carboxylic acids have a pH less than 7 and have the chemical properties typical of an acid.

Esters

Esters are a group of compounds formed when **carboxylic acids** and **alcohols** are warmed together in the presence of a catalyst, such as concentrated sulfuric acid. This can be carried out in a test tube in the laboratory.

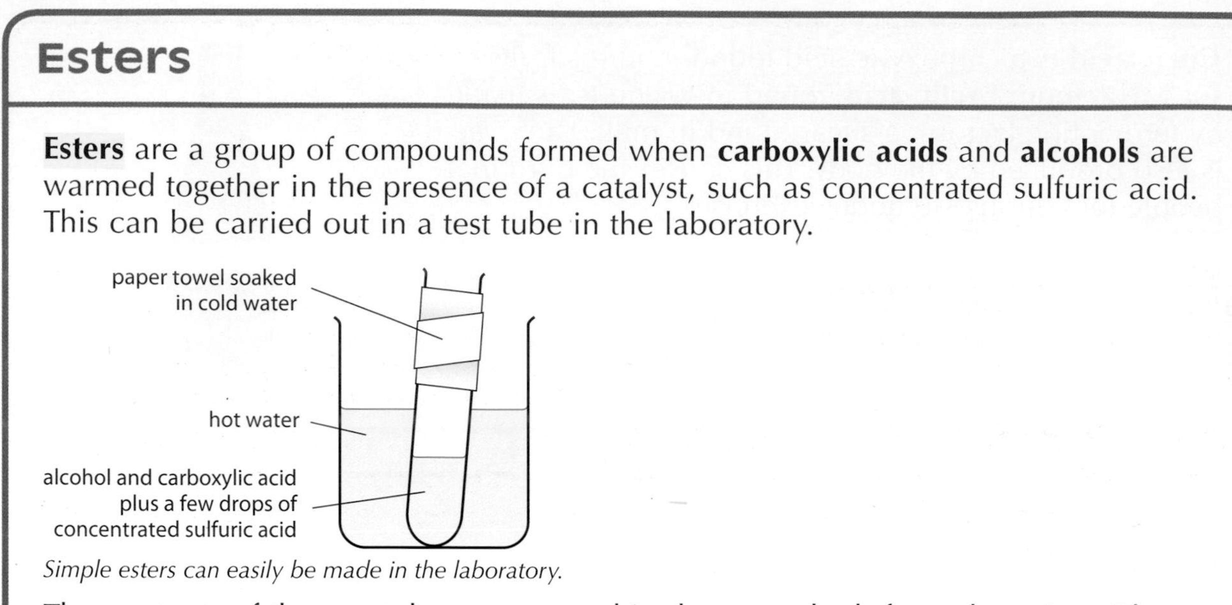

Simple esters can easily be made in the laboratory.

The contents of the test tube are warmed in the water bath for a short time. The paper towel soaked in cold water condenses the ester and stops it evaporating into the air. The reaction mixture is poured into cold water and an **oily layer** forms on top of the water – this is the ester. Esters have a characteristic **fruity smell**.

Naturally occurring esters are responsible for many flower scents and fruit flavours. Artificial esters are used in some confectionery products to give fruit flavours. Different combinations of alcohols and carboxylic acids give esters with different smells and flavours, as shown in the table – there is usually more than one ester responsible for a particular flavour.

Ester	Flavour
ethyl methanoate	raspberry
pentyl ethanoate	banana
propyl ethanoate	pear

The flavour in fruit is due to the presence of esters

Esters are named by combining the names of the alcohol and carboxylic acid: ethyl methanoate is made from ethanol and methanoic acid.

Esters are also good solvents for varnishes and paints. Nail varnishes use butyl ethanoate and ethyl ethanoate as solvents. Nail varnish remover also contains esters. The esters evaporate easily leaving behind the varnish. Esters can also be used as paint thinners for car paint.

Esters are used as solvents for nail varnish.

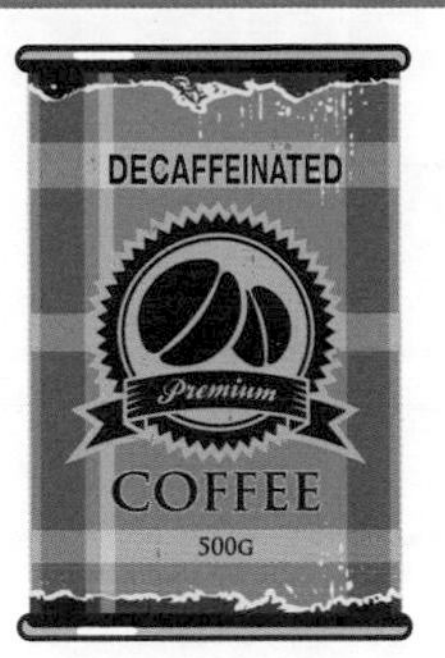

Ethyl ethanoate can be used as a solvent to take the caffeine out of coffee.

One of the methods that can be used to decaffeinate coffee uses ethyl ethanoate. Coffee beans are steamed and then soaked in the ethyl ethanoate, which dissolves out the caffeine. The beans are then steamed again to remove traces of the solvent, dried and roasted to give decaffeinated coffee.

Quick Test

1. The structural formula for a carboxylic acid is shown:

```
    H   H   H    O
    |   |   |   //
H — C — C — C = C
    |   |   |    \
    H   H   H     O — H
```

(a) Name the carboxylic acid.

(b) Write the molecular formula for the carboxylic acid.

2. Propanoic acid is a carboxylic acid.

(a) Write the molecular formula for propanoic acid.

(b) Draw the structural formula for propanoic acid.

(c) Propanoic acid is very miscible with water but octanoic acid has very low solubility. Explain why this is the case.

3. Name the first two carboxylic acids and give a use for each.

4. The word equation summarises the formation of an ester.

$$X + Y \xrightarrow[\text{heat}]{Z} \text{ester} + \text{water}$$

(a) Name X and Y and Z.

(b) What role does Z play in the reaction?

(c) If an ester is made in the laboratory, what two things would indicate that the ester has been formed?

(d) State two uses for esters.

Energy and chemicals from fuels

Calculating energy

What is actually being measured is the amount of heat transferred to the water but the procedure can be used to compare the energy produced by different fuels.

When fuels burn they give out energy – an **exothermic** reaction takes place. The fuel reacts with oxygen from the air – this is **combustion**.

$$C_2H_5OH(\ell) + 3O_2(g) \rightarrow 2CO_2(g) + 3H_2O(\ell)$$
ethanol

$$CH_4(g) + 2O_2(g) \rightarrow CO_2(g) + 2H_2O(\ell)$$
methane

When energy is taken in from the surroundings during a chemical reaction an **endothermic** reaction is said to take place.

The energy given out when a fuel burns can be calculated experimentally using a **calorimeter**. Industrial calorimeters can give very accurate measurements of the energy produced. In the lab a simple calorimeter can be made but it is not very accurate, mainly because of heat loss to the surroundings and the copper can.

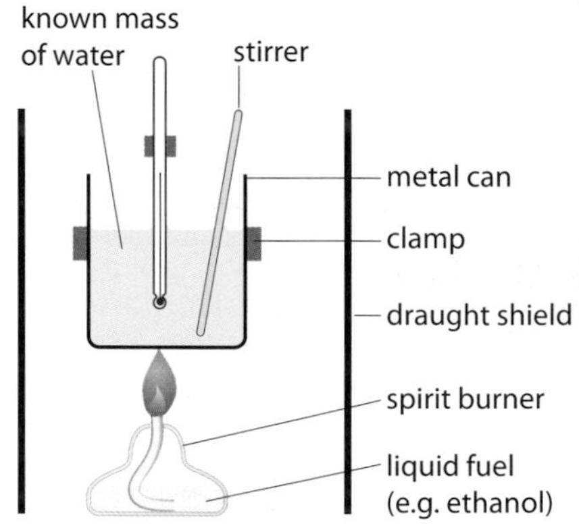

Simple laboratory calorimeter.

The energy given out (E_h) when a measured mass of fuel burns and heats a known mass of water (m) can be calculated by measuring the temperature rise of the water (ΔT) and using the equation:

E_h = c m ΔT (**c** is the specific heat capacity of water = 4.18 kJ kg^{-1} $°C^{-1}$)

Example: Calculate the energy transferred to the water when 1.0 g of ethanol burns, given the following experimental data:

Mass of water = 300 g (0.300 kg) (1 cm^3 of water = 1 g = 0.001 kg)
Temperature of water at the start = 20°C Temperature of water at the end = 44°C

Worked answer:
c = 4.18 m = 0.300 ΔT = 44 – 20 = 24

Substituting values into the equation: **E_h = c m ΔT** = 4.18 × 0.300 × 24 **E_h = 30.10 kJ**

Because of the loss of energy to the surroundings experimental values are always much lower than data booklet values.

If 1 g quantities of different fuels are burned in the same way they can be compared to see which gives out most energy.

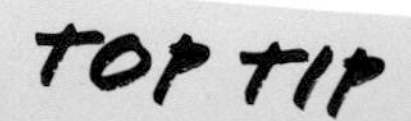

Energy is measured in kilojoules (kJ). The specific heat capacity of water (c) has complicated-looking units but they are not needed for the calculation.

Calculating masses reacting and produced

As well as comparing the amount of energy given out when different fuels burn, the mass of carbon dioxide given off when a fuel burns and the mass of oxygen required to completely burn a fuel can be calculated.

TOP TIP

One mole (1 mol) of a substance is the same as the gram formula mass (gfm). Look back at **Gram formula mass and the mole** before looking at calculating masses reacting and produced from balanced equations.

Example 1: Calculate the mass of carbon dioxide given off when 1 g of methanol is burned.

Balanced equation: $2CH_3OH(\ell) + 3O_2(g) \rightarrow 2CO_2(g) + 4H_2O(\ell)$

Worked answer: 2 mol ... 2 mol

So, 1 mol $\rightarrow$ 1 mol

Changing to grams 32 g (gfm) $\rightarrow$ 44 g (gfm)

$1\text{ g}\left(\frac{32}{32}\right) \rightarrow \frac{44}{32} = \mathbf{1.38\ g}$

Example 2: Calculate the mass of oxygen required to burn 50 g of methane.

Balanced equation: $CH_4(g) + 2O_2(g) \rightarrow CO_2(g) + 2H_2O(\ell)$

Worked answer: 1 mol ... 2 mol

Changing to grams 16 g needs 64 g

$1\text{ g}\left(\frac{16}{16}\right)$ needs $\frac{64}{16}$ g

So, 50 g needs $\frac{64}{16} \times 50 = \mathbf{200\ g}$

Quick Test

1. The experimental results below are obtained when a group of students burn 1 g of methanol (CH_3OH) in a simple laboratory calorimeter.

   Volume of water = 100 cm^3

   Temperature of water at the start = 20°C Temperature of water at the end = 53.5°C

   (a) Calculate the energy transferred to the water.

   (b) The energy transferred in the laboratory experiment carried out by the students is much less than the values found in the data book. Suggest why this might be the case.

   (c) Look at the diagram of the simple calorimeter. Explain why a draught shield is used and a copper beaker rather than a glass one.

2. Calculate the mass of carbon dioxide produced when 50 g of ethanol (C_2H_5OH) is burned completely.

   Balanced equation: $C_2H_5OH(\ell) + 3O_2(g) \rightarrow 2CO_2(g) + 3H_2O(\ell)$

Unit 2: Learning checklist

In this chapter you have learned:

- a homologous series is a group of compounds with similar chemical properties and physical properties that show a gradual change and can be represented by a general formula
- isomers are compounds with the same molecular formula but different structural formulae

Alkanes

- the alkanes are a homologous series of hydrocarbons with general formula C_nH_{2n+2}, where n = 1, 2, 3, etc.
- alkanes have branched-chain isomers
- alkanes are generally unreactive but burn and are used as fuels

Naming and drawing branched alkanes

- how to name branched-chain alkanes, given their structure, and draw structural formulae, given the name of an alkane

Cycloalkanes

- cycloalkanes are a homologous series of saturated hydrocarbons where the carbons are joined in a ring and have a general formula C_nH_{2n} where n = 3, 4, etc.
- how to name cycloalkanes
- how to draw isomers of the cycloalkanes
- the cycloalkanes are present in petrol and are used as solvents

Alkenes

- that alkenes are a homologous series of hydrocarbons with general formula C_nH_{2n}, where n = 2, 3, 4, etc.
- alkenes have straight-chain and branched-chain isomers
- how to name alkenes systematically

Reactions of alkenes

- the C=C double bond in alkenes makes them more reactive than alkanes
- the alkenes can undergo addition reactions when other molecules join by adding across the double bond
- hydrogenation and hydration are two examples of addition reactions with important industrial applications
- Alkenes decolorise bromine solution very quickly in an addition reaction because alkenes are unsaturated, while alkanes only decolorise bromine solution very slowly because they are saturated and do not undergo addition reactions

Alcohols

- that alcohols are a homologous series of compounds containing the hydroxyl (–OH) functional group
- the general formula for the alcohols is $C_nH_{2n+1}OH$, where n = 1, 2, 3, etc.
- to write molecular formulae and draw structural formulae for straight-chain alcohols, given their systematic name
- to systematiclly name straight-chain alcohols, given their shortened or full structural formulae

Properties and reactions of alcohols

- alcohols are very good solvents
- alcohols burn with a very clean flame and can be used as fuels

Carboxylic acids and esters

- carboxylic acids are a homologous series of compounds containing the carboxyl (–COOH) functional group
- the general formula for the alcohols is $C_nH_{2n+1}COOH$, where n = 0, 1, 2, etc.
- to write molecular formulae and draw structural formulae for straight-chain carboxylic acids, given their name
- to name straight-chain carboxylic acids, given their shortened or full structural formulae
- vinegar is a solution of ethanoic acid in water
- vinegar can be used as a preservative
- esters are made by reacting alcohols with carboxylic acids
- esters are used as food flavourings and as solvents

Energy and chemicals from fuels

- combustion, burning of fuels, is an exothermic process – heat energy is given out
- endothermic reactions take in energy from their surroundings
- fuels can be compared by measuring the energy given out when they are burned
- when a burning fuel is used to heat water the energy transferred is calculated using the formula $E_h = cm\Delta T$
- balanced equations can be used to calculate the masses of reactants and products of a reaction
- carry out calculations using balanced equations

Metals from ores

Percentage composition by mass

An **ore** is a type of rock that contains important elements including metals. Some common ores that metals can be extracted from are haematite (Fe_2O_3), chalcocite (Cu_2S) and cinnebar (HgS).

The % by mass of a metal in a compound can be calculated from its formula.

Example: Cu_2S

$$\text{\% composition by mass of Cu} = \frac{\text{mass of Cu in formula}}{\text{formula mass of } Cu_2S} \times 100$$

$$= \frac{(2 \times 63.5)}{(2 \times 63.5) + (1 \times 32)} \times 100$$

$$= \frac{127}{159} \times 100$$

% composition by mass of Cu = 79.9%

Extraction of metals from their ores

Heating

Some metals, like gold, can be found uncombined in the Earth's crust but most metals exist as compounds in an ore.

Metals **low in reactivity**, such as silver, can be obtained simply by **heating** their ore. Ores often contain metal oxides.

If silver(I) oxide is heated, silver metal is produced:

$$2Ag_2O(s) \rightarrow 4Ag(s) + O_2(s)$$

The metal ore is said to have been **reduced** to the metal.

During the reaction the positive silver ions change to atoms by gaining an electron. This can be seen in the ionic equation:

$$2(Ag^+)_2O^{2-}(s) \rightarrow 4Ag(s) + O_2(s)$$

The positive silver ion has gained an electron and formed the metal atom. Gain of electrons is called **reduction**.

This can be shown in an **ion-electron equation**. Ion-electron equations show the electrons gained or lost by an atom or ion.

$$Ag^+(s) + e^- \rightarrow Ag(s) \qquad \text{reduction}$$

TOP TIP

The SQA data booklet gives a list of commonly used ion-electron equations.

Smelting

Some ores that are too reactive to be reduced by heat alone can be reduced by heating with carbon. This process is called **smelting**.

One of the most important examples of smelting is the extraction of iron from iron ore in a blast furnace. The diagram summarises what happens in the blast furnace. The coke loaded in at the top provides the carbon for the reaction.

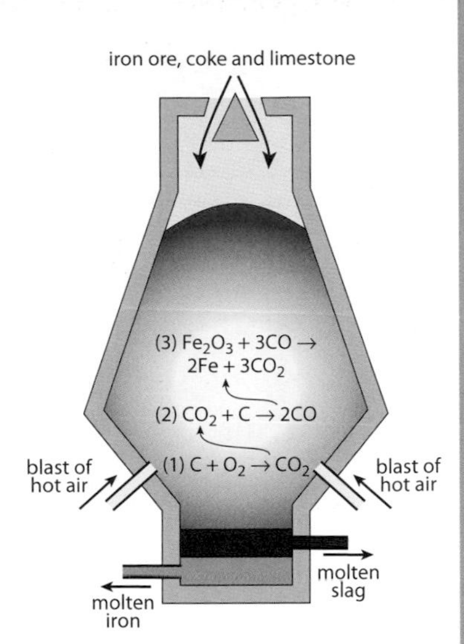

Iron is extracted from its ore in a blast furnace.

The ionic equation for the reduction of iron(III) oxide shows the Fe^{3+} ion being **reduced** to Fe:

$$(Fe^{3+})_2(O^{2-})_3(s) + 3CO(g) \rightarrow 2Fe(\ell) + 3CO_2(g)$$

The ion-electron equation can be written for the reduction of the iron ions:

$$Fe^{3+}(s) + 3e^- \rightarrow Fe(\ell) \quad \text{reduction}$$

The carbon monoxide causes the reduction so is known as the **reducing agent**.

Reducing agents give up electrons allowing substances being reduced to gain electrons.

Using electricity

Some metals are too reactive to be reduced by chemical means. Electricity has to be used. A **direct current (dc)**, which gives a positive and negative electrode, has to be used so that the products can be identified. The positive ions are attracted to the negative electrode where they gain electrons, i.e. they are reduced. This process is known as **electrolysis**.

Aluminium is an important metal extracted from its oxide (ore) by passing a dc current through it:

$$Al^{3+}(\ell) + 3e^- \rightarrow Al(\ell) \quad \text{reduction}$$

TOP TIP

The electrochemical series in the SQA data booklet will help you to decide which method is needed to extract the metal from its compound. Metals below copper can be extracted by heat alone. Metals above zinc need electricity. Metals in the middle can be reduced by heating with carbon.

Quick Test

1. Calculate the percentage composition by mass of mercury in mercury sulfide (HgS).
2. Copper can be extracted from its oxide by heating the oxide and passing carbon monoxide over it. The equation is shown:

   $CuO(s) + CO(g) \rightarrow Cu(s) + CO_2(g)$

   (a) What name is given to the process of extracting a metal in this way?
   (b) Write the ion-electron equation to show what happens to the copper(II) ion.
   (c) Identify the reducing agent.

Properties and reactions of metals

Remember from National 4!

Metals have a number of properties in common:

- are good conductors of heat and electricity
- tend to be shiny – have metallic lustre
- are malleable – can be shaped
- are ductile – can be stretched into wires.

Metallic bonding

The properties of metals can be explained by the bonding within the metals.

In metals the outer electrons of the atoms can move easily from atom to atom. The electrons within the structure are said to be **delocalised**. Metal structures can be described as 'positive ions in a sea of electrons'.

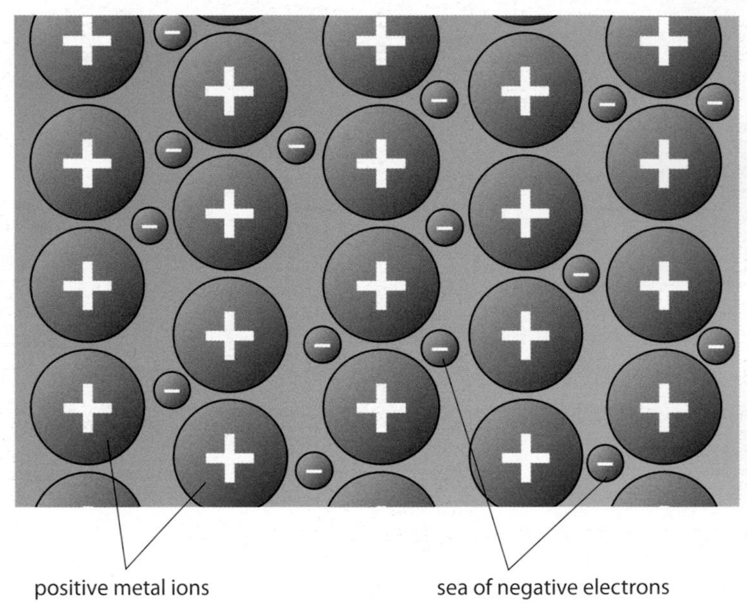

Metals exist as positive ions in a 'sea' of delocalised electrons.

The structure is held together by metallic bonds. These are the attractive forces of the metal nuclei for the delocalised electrons moving between them. The direction of the bonds is not fixed because the electrons are moving, unlike in covalent compounds where the electrons are localised in bonds. This means that the atoms will be able to move in relation to each other and explains why the metals can be rolled into thin sheets or drawn into wires.

Metals are good electrical conductors because the delocalised electrons can move through the metal structure, i.e. an electrical current will flow through the metal.

Reactions of metals: redox reactions

The reactions of metals with oxygen, water and dilute acid can be used to work out an order of reactivity of metals known as the reactivity series.

TOP TIP

Ion-electron equations are found in the SQA data booklet. Reductions are written as they are in the booklet, and oxidations are reversed.

Metals and oxygen

When metals react with oxygen the metal oxide is produced.

Magnesium reacts violently with oxygen to form magnesium oxide:

$$2Mg(s) + O_2(g) \rightarrow 2MgO(s)$$

Ionic equation: $2Mg(s) + O_2(g) \rightarrow 2Mg^{2+}O^{2-}(s)$

Magnesium reacts violently with the oxygen in the air to form magnesium oxide.

The ionic equation shows the magnesium atoms have **lost electrons** and formed magnesium ions. Loss of electrons is called **oxidation**. The ion-electron equation for oxidation is written as:

$Mg(s) \rightarrow Mg^{2+}(s) + 2e^-$ **oxidation**

The oxygen has gained electrons from the magnesium so has been **reduced:**

$O_2(g) + 4e^- \rightarrow 2O^{2-}(s)$ **reduction**

Reduction and **ox**idation always happen together in a **redox** reaction.

The reduction and oxidation ion-electron equations can be added together in a redox equation. Electrons are not shown in a redox reaction because the electrons lost by the magnesium are gained by the oxygen, so they cancel out. The ion-electron equations are multiplied up to balance the electrons lost and gained:

Reduction: $O_2(g) + 4e^- \rightarrow 2O^{2-}(s)$

Oxidation: $2Mg(s) \rightarrow 2Mg^{2+}(s) + 4e^-$ (equation multiplied by 2 to balance the electrons)

Add to get

redox equation: $2Mg(s) + O_2(g) \rightarrow 2Mg^{2+}O^{2-}(s)$
(electrons cancel out)

TOP TIP

Ion-electron equations sometimes need to be multiplied to balance the electrons before adding to make the redox equation.

Metals and water

Sodium reacts violently with water to form sodium hydroxide.

Alkali metals (Group 1 in the periodic table) are so called because they react with water to give alkaline solutions.

When sodium reacts with water, sodium hydroxide and hydrogen gas are produced:

$$2Na(s) + 2H_2O(\ell) \rightarrow 2NaOH(aq) + H_2(g)$$

Ionic equation:

$$2Na(s) + 2H_2O(\ell) \rightarrow 2Na^+(aq) + 2OH^-(aq) + H_2(g)$$

The ionic equation shows the sodium has been oxidised and the water reduced:

oxidation $Na(s) \rightarrow Na^+(aq) + e^-$

reduction $2H_2O(\ell) + 2e^- \rightarrow H_2(g) + 2OH^-(aq)$

redox $2Na(s) + 2H_2O(\ell) \rightarrow 2Na^+(aq) + 2OH^-(aq) + H_2(g)$

Metals and acids

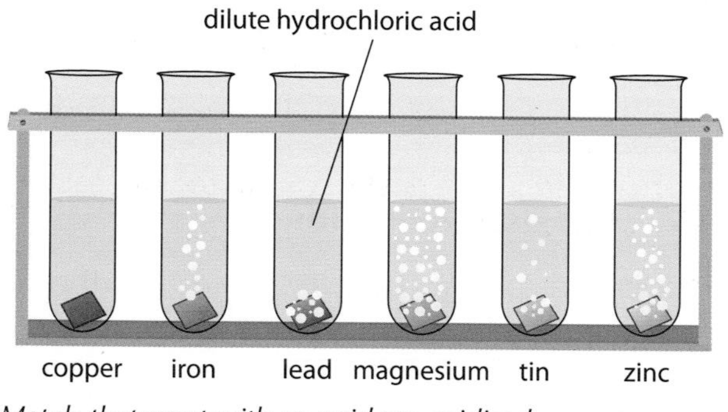

Metals that react with an acid are oxidised.

When zinc reacts with dilute hydrochloric acid, zinc chloride and hydrogen are produced:

$$Zn(s) + 2HCl(aq) \rightarrow ZnCl_2(aq) + H_2(g)$$

Ionic equation: $Zn(s) + 2H^+(aq) + 2Cl^-(aq) \rightarrow Zn^{2+}(aq) + 2Cl^-(aq) + H_2(g)$

The ionic equation shows the zinc atoms have been oxidised and the hydrogen ions reduced:

oxidation	$Zn(s)$	$\rightarrow$	$Zn^{2+}(aq)$	+	$2e^-$		
reduction	$2H^+(aq)$	+	$2e^-$	$\rightarrow$	$H_2(g)$		
redox	$Zn(s)$	+	$2H^+(aq)$	$\rightarrow$	$Zn^{2+}(aq)$	+	$H_2(g)$

Ion-electron and redox equations do not show ions that do not react (spectator ions). In this example the Cl^- ion is a spectator ion.

TOP TIP

You can remember the meaning of oxidation and reduction by the phrase 'OIL RIG'

Oxidation **I**s **L**oss of e^-

Reduction **I**s **G**ain of e^-

Quick Test

1. Copper metal is used in electrical flexes because it is a good conductor of electricity. Explain what makes metals good conductors of electricity.
2. Zinc reacts with oxygen to form zinc oxide:

   $2Zn(s) + O_2(g) \rightarrow 2ZnO(s)$

   (a) Write the ionic equation for the reaction.

   (b) Write the ion-electron equation for the oxidation of zinc.

   (c) Write the ion-electron equation for the reduction of oxygen.

   (d) Write the redox equation for the reaction.
3. When lithium reacts with water, lithium hydroxide and hydrogen gas are produced:

   $2Li(s) + 2H_2O(\ell) \rightarrow 2LiOH(aq) + H_2(g)$

   (a) Write the ionic equation for the reaction.

   (b) Write the ion-electron equation for the oxidation of lithium.

   (c) Write the ion-electron equation for the reduction of water.

   (d) Write the redox equation for the reaction.
4. When magnesium reacts with dilute sulfuric acid, magnesium sulfate and hydrogen are produced:

   $Mg(s) + H_2SO_4(aq) \rightarrow MgSO_4(aq) + H_2(g)$

   (a) Write the ionic equation for the reaction.

   (b) Write the ion-electron equation for the oxidation of magnesium.

   (c) Write the ion-electron equation for the reduction of the hydrogen ions.

   (d) Write the redox equation for the reaction.

Electrochemical cells

Remember from National 4!

When different metals are connected and placed in an electrolyte, an electric current flows. This is called an electrochemical cell. Comparing voltage readings between pairs of metals can be used to construct an electrochemical series. The electrochemical series can be used to predict the voltage and current direction in an electrochemical cell. A battery is a type of electrochemical cell.

Cells involving metals

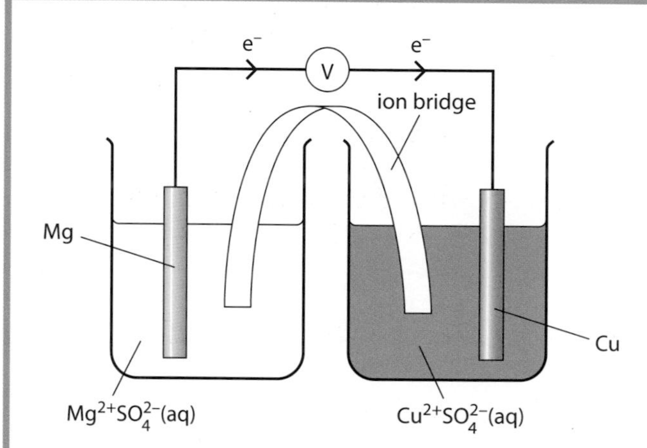

A chemical reaction takes place at each electrode in a chemical cell.

The **ion bridge** contains an electrolyte that allows ions to flow between the two solutions. Electrons flow from magnesium to copper, through the wire, because magnesium is higher than copper in the electrochemical series.

At the **magnesium electrode**: electrons flow from Mg → Cu (through the wire), so magnesium is **losing electrons** – magnesium atoms are being **oxidised** to magnesium ions. The ion-electron equation is:

oxidation $Mg(s) \rightarrow Mg^{2+}(aq) + 2e^-$

At the **copper electrode**: the copper ions in the solution **gain electrons**, which have come from the magnesium. The copper ions are **reduced** to copper atoms. The ion-electron equation is:

reduction $Cu^{2+}(aq) + 2e^- \rightarrow Cu(s)$

Adding the reduction and oxidation equations gives the **redox** equation. Notice here that the electrons balance so there is no need to multiply either of the ion-electron equations. The spectator ion, ($SO_4^{2-}(aq)$), is not included.

redox $Mg(s) + Cu^{2+}(aq) \rightarrow Mg^{2+}(aq) + Cu(s)$

TOP TIP

Read about reduction, oxidation and redox equations in **Properties and reactions of metals** before doing this section.

Cells involving non-metals

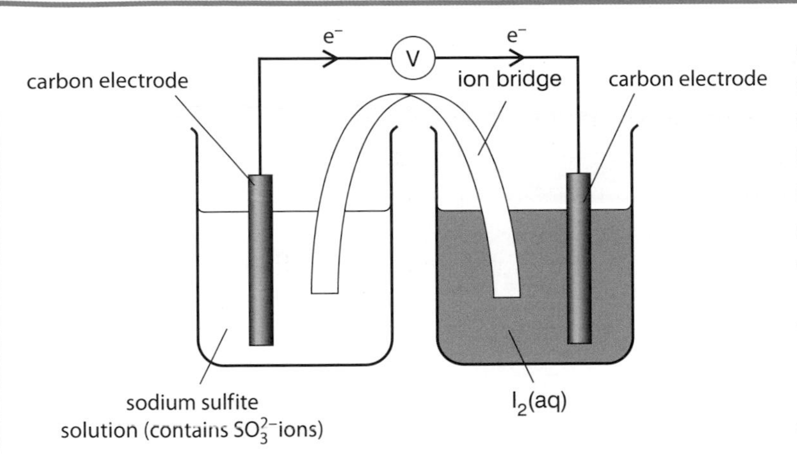

A chemical cell can be set up using non-metal electrodes. **Graphite** (a form of carbon) is often used because it conducts electricity and will not react with the electrolytes.

An example is the reaction between sodium sulfite and iodine:

The electrons flow from left to right. This is because at the left-hand electrode the sulfite ion is losing electrons – it is being **oxidised**:

oxidation $SO_3^{2-}(aq) + H_2O(\ell) \rightarrow SO_4^{2-}(aq) + 2H^+(aq) + 2e^-$

At the right-hand electrode the iodine molecule is gaining electrons – it is being **reduced**:

reduction $I_2(aq) + 2e^- \rightarrow 2I^-(aq)$

redox $SO_3^{2-}(aq) + H_2O(\ell) + I_2(aq) \rightarrow SO_4^{2-}(aq) + 2I^-(aq) + 2H^+(aq)$

Quick Test

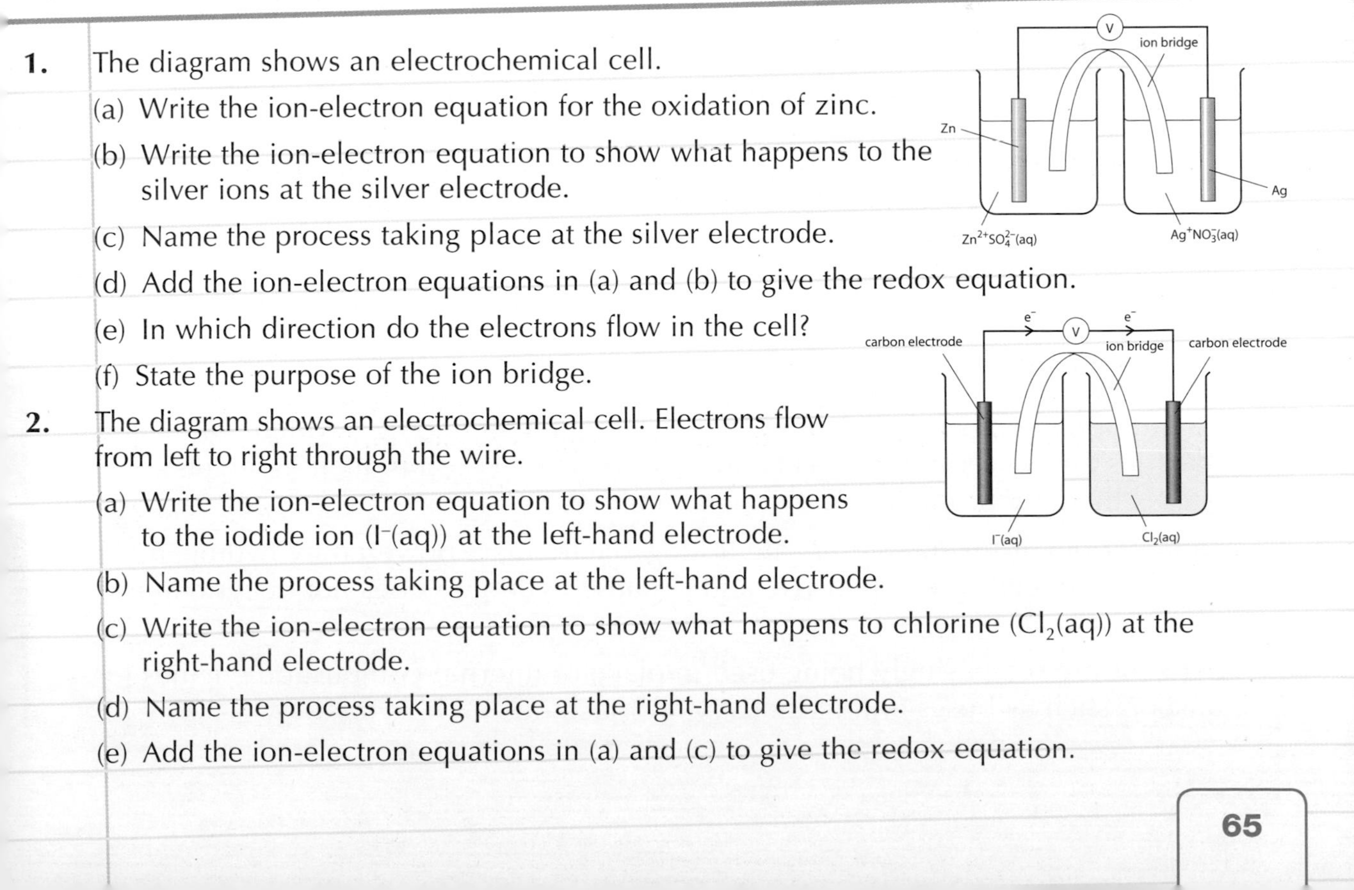

1. The diagram shows an electrochemical cell.
   (a) Write the ion-electron equation for the oxidation of zinc.
   (b) Write the ion-electron equation to show what happens to the silver ions at the silver electrode.
   (c) Name the process taking place at the silver electrode.
   (d) Add the ion-electron equations in (a) and (b) to give the redox equation.
   (e) In which direction do the electrons flow in the cell?
   (f) State the purpose of the ion bridge.
2. The diagram shows an electrochemical cell. Electrons flow from left to right through the wire.
   (a) Write the ion-electron equation to show what happens to the iodide ion ($I^-(aq)$) at the left-hand electrode.
   (b) Name the process taking place at the left-hand electrode.
   (c) Write the ion-electron equation to show what happens to chlorine ($Cl_2(aq)$) at the right-hand electrode.
   (d) Name the process taking place at the right-hand electrode.
   (e) Add the ion-electron equations in (a) and (c) to give the redox equation.

Technologies that use redox reactions

Fuel cells

A **fuel cell** is an electrochemical cell that produces electricity by combining fuels such as hydrogen and alcohol with oxygen without burning them.

A **hydrogen** fuel cell consists of two electrodes with a special membrane called a **proton exchange membrane (PEM)**, which acts as the electrolyte in the cell but allows only protons (**H^+ ions**) to move through it.

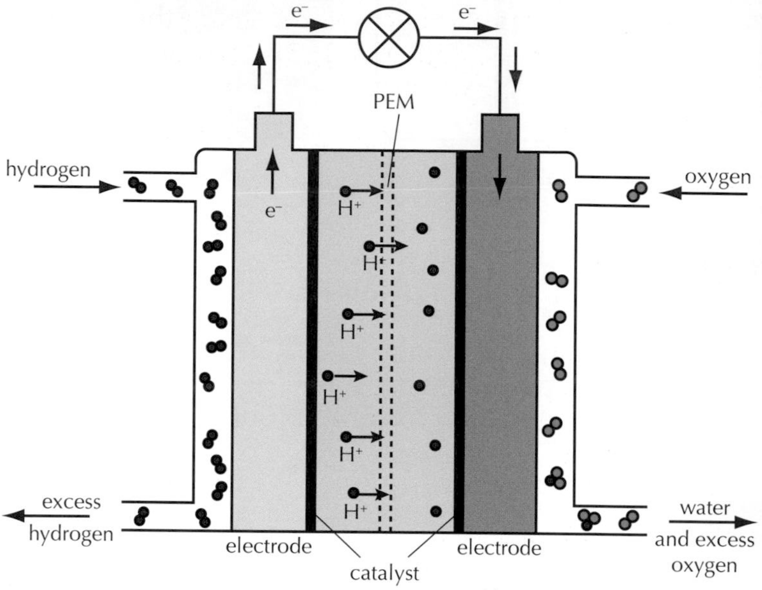

Diagram of a hydrogen fuel cell.

At the left-hand electrode hydrogen reacts with a **catalyst**, creating positively charged hydrogen ions and electrons. The ions then pass through the membrane while the electrons (e^-) move through a wire and light the bulb or power any other machine attached to the cell. At the right-hand electrode oxygen reacts with the hydrogen ions and electrons forming water. Several cells are usually linked together to increase the voltage produced.

The ion-electron equations for the oxidation and reduction reactions in the cell are:

oxidation $H_2(g) \rightarrow 2H^+(aq) + 2e^-$

reduction $O_2(g) + 4H^+(aq) + 4e^- \rightarrow 2H_2O(\ell)$

redox $2H_2(g) + O_2(g) \rightarrow 2H_2O(\ell)$

The oxidation equation is multiplied by 2 to balance the electrons in the reduction equation.

Remember!

The use of fuel cells is increasing. They are a clean technology, which will help industries and governments meet carbon dioxide emission reduction targets – water is the only product in a hydrogen fuel cell.

A number of car manufacturers are testing electric cars powered by hydrogen fuel cells, but fuel cells are expensive, and because hydrogen does not occur naturally, it has to be made.

Fuel cells are increasingly being used in place of internal combustion engines for transport such as buses and in industry for machines such as forklifts.

Rechargeable batteries

Rechargeable batteries are batteries which when they go 'flat' can be recharged and re-used. Life would be very different without rechargeable batteries; laptop computers, mobile phones and electric cars are all powered by rechargeable batteries.

Rechargeable batteries, like all batteries, use redox reactions to produce electrical currents. Recharging involves reversing the redox reaction, which produces electricity. The original state of the battery is regenerated, allowing the battery to produce more electricity.

A lead-acid battery is a rechargeable battery used in cars. It consists of lead plates, some of which are coated with lead(IV) oxide.

During **discharge**, i.e. when the battery is producing electricity, the reactions taking place at the electrodes are:

$Pb(s) + SO_4^{2-}(aq) \rightarrow PbSO_4(s) + 2e^-$
oxidation

$PbO_2(s) + 4H^+(aq) + SO_4^{2-}(aq) + 2e^- \rightarrow PbSO_4(s) + 2H_2O(\ell)$
reduction

The overall redox equation for the **discharging** reaction in the battery is:

$Pb(s) + PbO_2(s) + 4H^+(aq) + 2SO_4^{2-}(aq) \rightarrow 2PbSO_4(s) + 2H_2O(\ell)$

During **recharging** the chemical reactions are **reversed**.

A range of very small rechargeable batteries has been developed for use in mobile phones and laptops.

Quick Test

1. Methanol fuel cells are being developed. The ion-electron equation for one of the reactions is:

$$CH_3OH(\ell) + H_2O(\ell) \rightarrow 6H^+(aq) + CO_2(g) + 6e^-$$

(a) State the name given to this type of ion-electron equation.

(b) State an environmental problem associated with using a methanol fuel cell.

2. Nickel-cadmium rechargeable batteries have been used in a number of portable devices. The ion-electron equations for the **discharge** of the battery are:

(i) $Cd + 2OH^- \rightarrow Cd(OH)_2 + 2e^-$

(ii) $NiOOH + H_2O + e^- \rightarrow Ni(OH)_2 + OH^-$

(a) Which of the ion-electron equations is reduction?

(b) Combine (i) and (ii) to form the redox equation.

(c) Write the ion-electron equations for the processes that take place when the battery is **recharged** and label them as reduction and oxidation.

Addition polymerisation

Remember from National 4!

Polymers are giant molecules formed when lots of small molecules, known as monomers, join together. Alkene monomers are made by cracking long-chain alkanes.

The **poly(ethene)** used to make the playing surface of the London 2012 Olympic hockey pitch is an example of an **addition polymer**. Addition polymers are formed when lots of small **unsaturated** molecules (**monomers**) join together – in the case of poly(ethene) the monomer is **ethene**. The characteristic that makes hydrocarbon molecules like ethene suitable to undergo an addition reaction is the carbon-to-carbon **double bond (C=C)**. Under the right conditions the double bond breaks, resulting in electrons being made available to bond with neighbouring ethene molecules. Although the diagram shows only three molecules joining, in practice thousands of molecules join together.

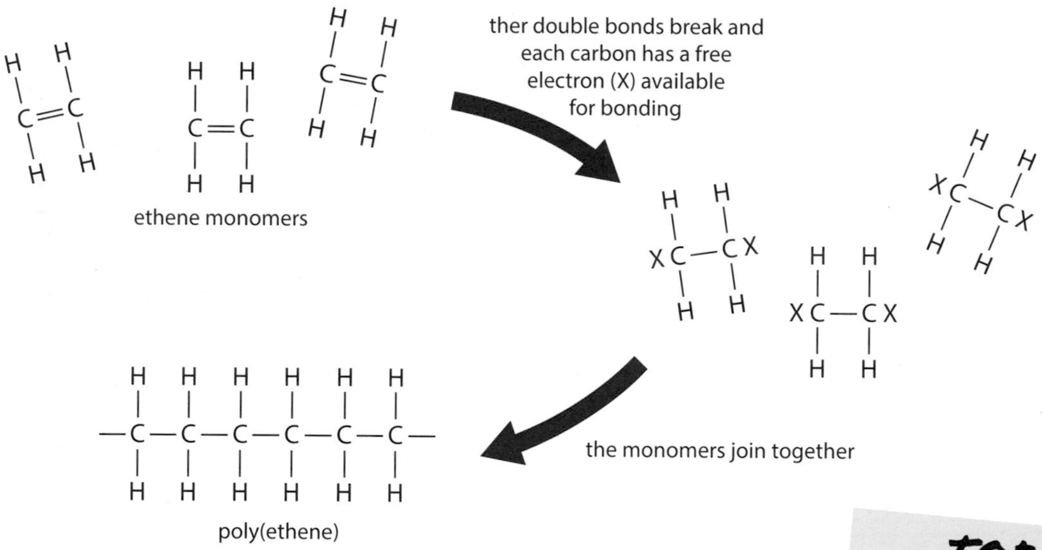

Lots of ethene molecules join to make poly(ethene).

Varying the number of carbon atoms in the polymer chain can change its properties. Poly(ethene) can be made soft and very flexible by keeping the chain lengths around 50 000 carbon atoms and having lots of branches. It can be made more rigid by making the chain length longer and having fewer branches.

TOP TIP

All addition polymers are made in the same way:

- The monomers must have a C=C double bond.
- The double bonds break.
- The monomers join to form a polymer.
- You can recognise an addition polymer because it has only carbon in its backbone.

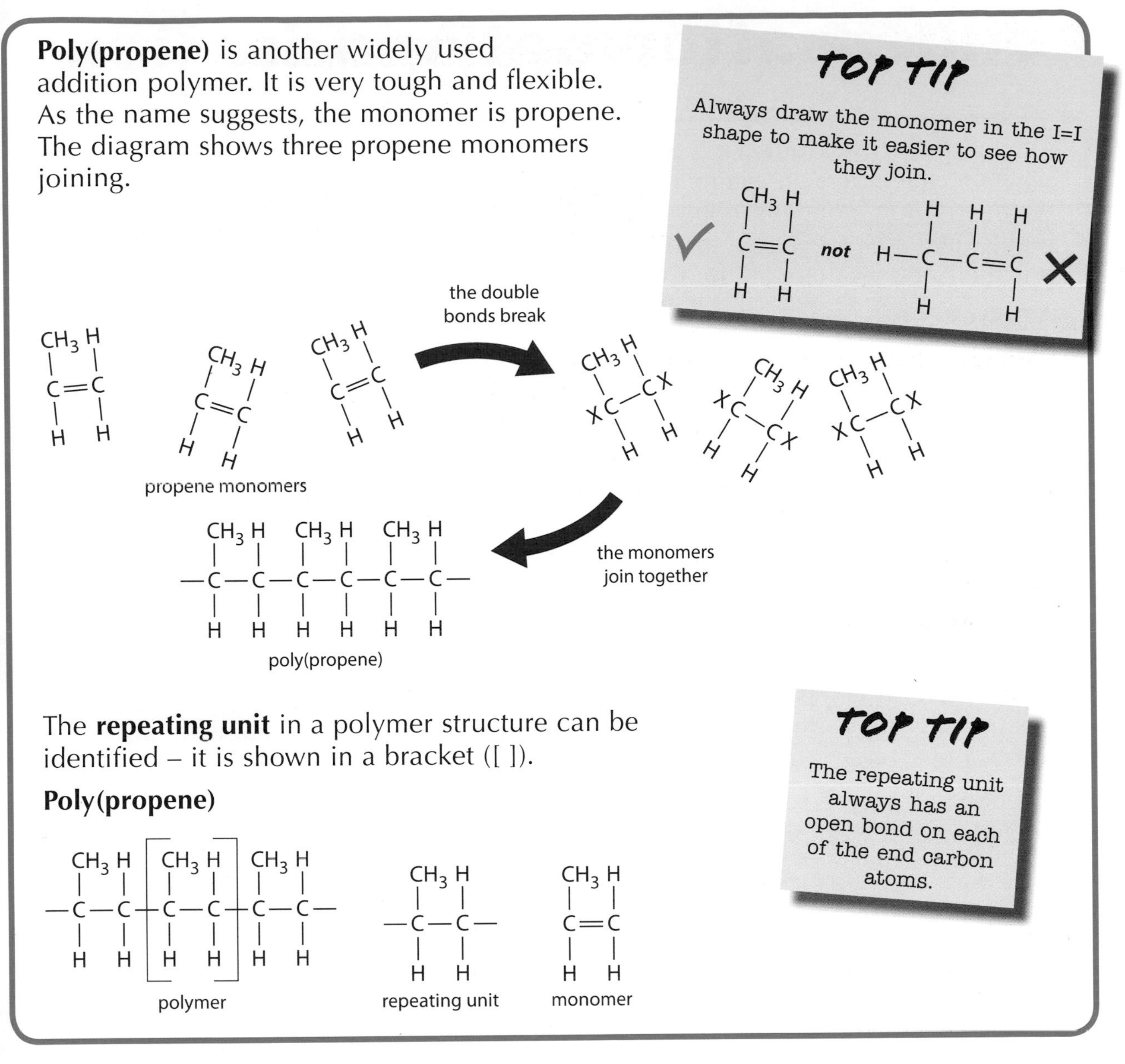

Quick Test

1. The monomer used to make polystyrene can be represented as:
   (a) Name the monomer.
   (b) Show how three of these monomer units join to form polystyrene.
   (c) Draw the repeating unit in polystyrene.
2. Part of the poly(tetrafluoroethene) molecule is shown:
   (a) Draw the repeating unit.
   (b) Draw the monomer unit.
   (c) Name the monomer.

Condensation polymerisation

The **polyester** used to make the panels that were used on the main London 2012 Olympic stadium is an example of a **condensation polymer**. Just like addition polymerisation, condensation polymers are formed when lots of monomers join together. However the monomers are different to the ones used in addition polymerisation – there are **no C=C** double bonds.

A polyester is formed when lots of monomers with hydroxyl and carboxyl groups join together. The monomers involved have to be **bifunctional**. This means that each monomer has to have two functional groups so that a condensation reaction can take place at each end of the monomer.

An alcohol with two –OH functional groups is often called a **diol**. A carboxylic acid with two –COOH functional groups is called a **dicarboxylic acid**.

H—O—□—O—H

a diol

H—O—C(=O)—⬭—C(=O)—O—H

a dicarboxylic acid

The square and oval parts in the structures indicate that it doesn't matter which group of atoms is in the middle of the molecule so long as they have the two functional groups. This is because the reaction occurs around the functional groups only.

The diol and dicarboxylic acid molecules react at both ends of the molecules producing a polyester.

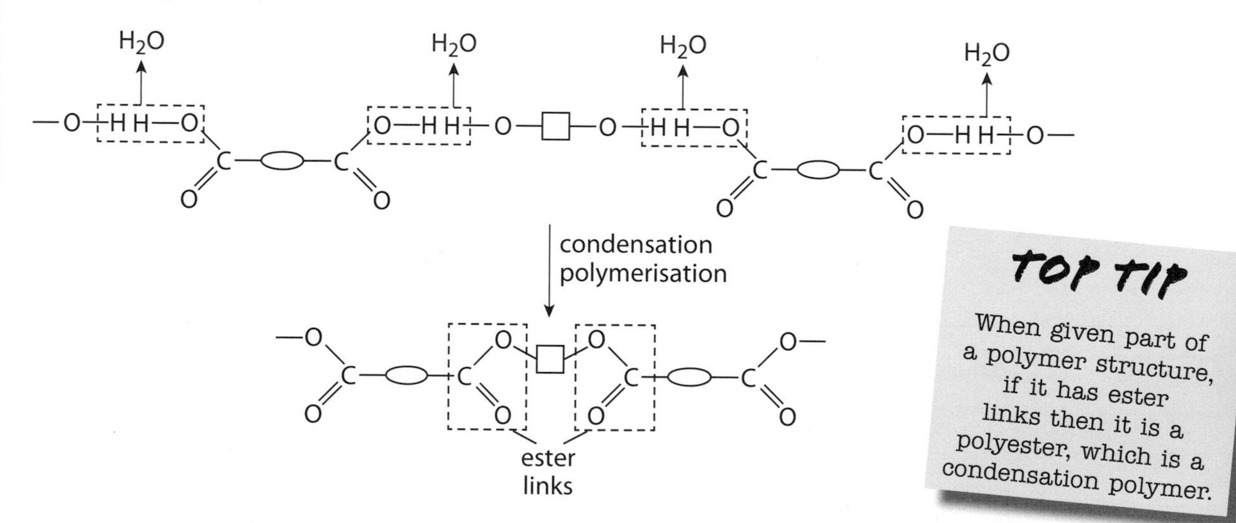

Using monomers with different functional groups produces other condensation polymers. **Nylon** is an example of a type of condensation polymer where one of the monomers used has $–NH_2$ (amino) groups at either end of the molecule that react with a dicarboxylic acid:

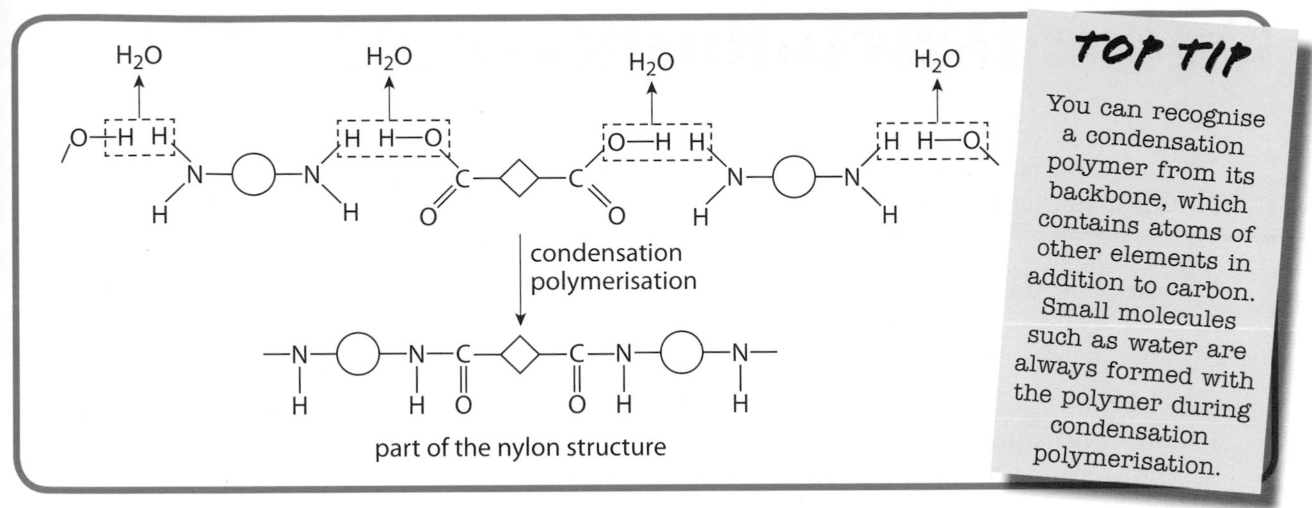

TOP TIP

You can recognise a condensation polymer from its backbone, which contains atoms of other elements in addition to carbon. Small molecules such as water are always formed with the polymer during condensation polymerisation.

Identifying the monomers

Identifying the structure of the monomers in condensation polymers means identifying two monomers. The first step is to identify the link between the monomers. This is where the structure splits.

For a polyester, it is the **opposite** of the formation process.

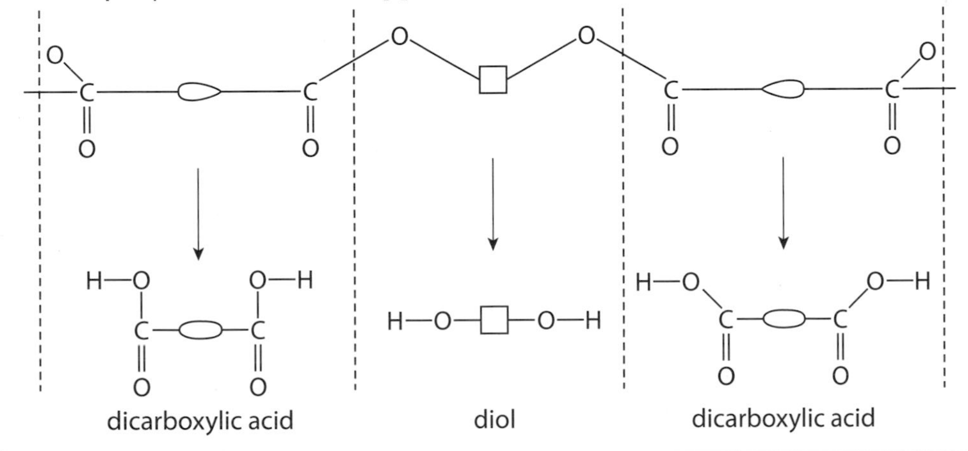

Quick Test

1. Three of the monomer units that join to form the condensation polymer terylene are shown below:

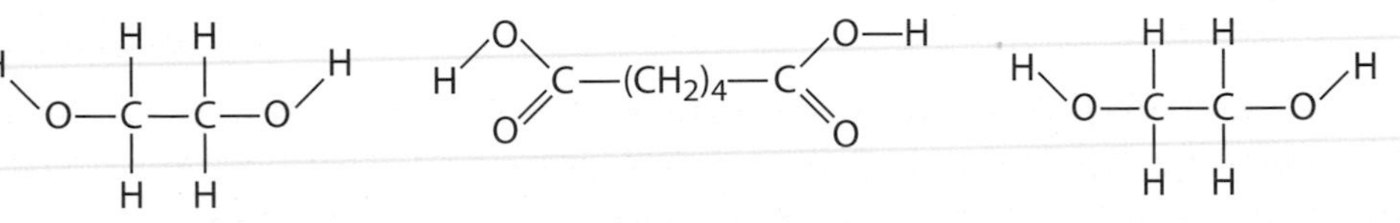

(a) (i) Draw the part of the terylene polymer formed when the three monomers join.

(ii) A small molecule is also formed when the monomer units join. Name the small molecule formed. Show how it is made, by circling together the elements involved in making the small molecule, on your diagram in part (i).

(b) What type of condensation polymer is terylene?

Natural polymers

Starch, cellulose, protein and DNA are all **natural condensation polymers**, sometimes called **biopolymers**, which are produced by living things.

Starch is made when lots of **glucose monomers** polymerise. The structure of the glucose is very complicated but can be represented as:

We can represent three glucose molecules joining together to form part of a starch molecule as:

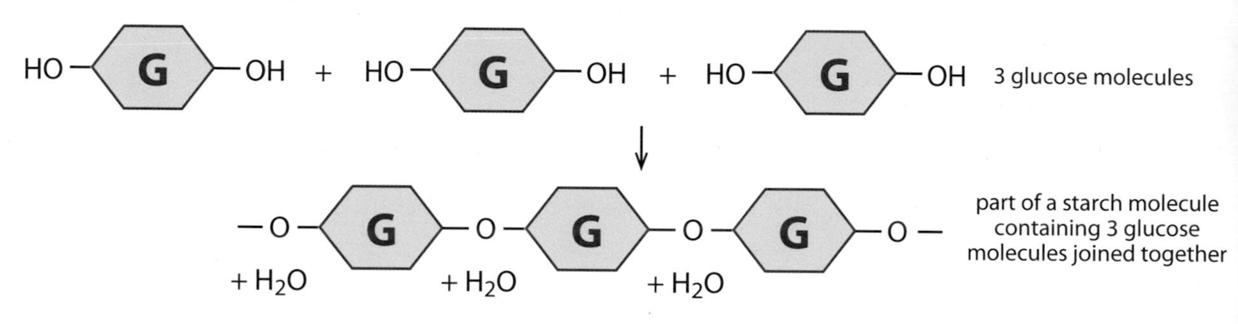

Note how water molecules are formed during the polymerisation, showing this is a **condensation** reaction.

Starch is an energy source for us when it is digested in the body. Starchy foods include potatoes, pasta and bread.

Cellulose is made when glucose molecules with a slightly different structure to the ones found in starch polymerise. Cellulose is probably the world's most abundant organic chemical. It is part of the structure of plants. We cannot digest cellulose because we don't have the necessary enzymes to digest it. However, cellulose gives us the fibre we require in our diet.

Proteins are the major structural material of animal tissue, e.g. muscles, skin, hair and nails. They are also involved in the maintenance and regulation of life processes. They are natural condensation polymers made from monomers called amino acids that have two functional groups – they are bifunctional.

DNA is a condensation polymer formed when monomers called nucleotides polymerise. DNA is found in almost all living things. It carries genetic information that can be passed from one generation to the next.

Natural rubber is a mixture of polyisoprene and small amounts of other chemicals including water. Rubber is often referred to as an elastomer (elastic polymer). Rubber is derived from latex, a milky substance produced by some plants. The main source of latex is the rubber tree, mainly grown in parts of Asia.

Bioplastics are synthetic and are a form of plastic derived from **renewable biomass** sources, such as vegetable fats and oils, corn starch and pea starch. They were the basis for the production of the earliest plastics and are making a comeback. The monomers used to make common plastics like poly(ethene) come mainly from oil fractions, which are becoming scarce – oil is a **finite** resource. There are a variety of bioplastics being made – some common uses are as packaging materials and insulation. Poly(lactic acid), known as PLA, is a **biodegradable** bioplastic – it breaks down naturally over time – which makes it more environmentally friendly. Waste sacks, disposable cutlery and even teabags can be made of PLA. The diagram below shows how three lactic acid monomers join to form part of the PLA structure:

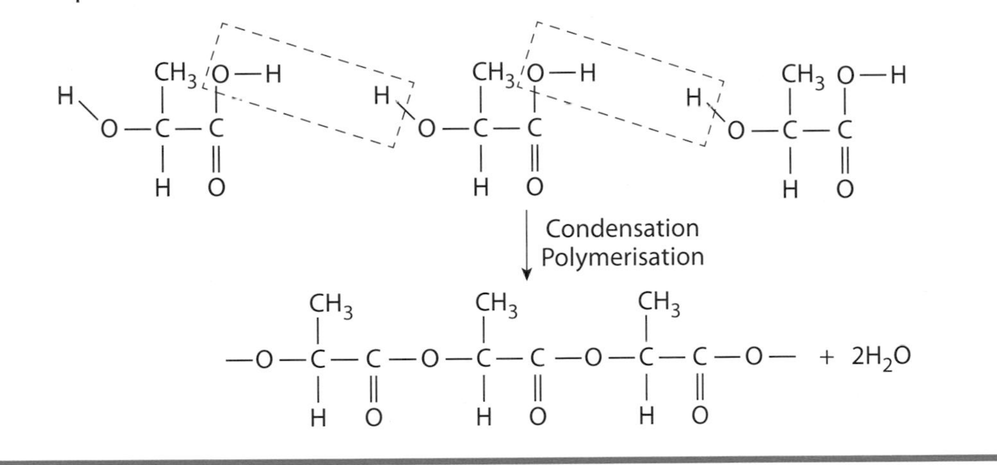

Quick Test

1. (a) Name the two functional groups present in lactic acid.
   (b) What term is used to describe a molecule with two functional groups?
   (c) What type of condensation polymer is poly(lactic acid)?
   (Hint: think about the link formed between the monomers.)
2. Complete the summary by filling in the blanks. You may wish to use the word bank to help you.
   Cellulose and starch are examples of (a) ______________ polymers formed by (b) ______________ polymerisation. They are also known as (c) ______________. The monomers that polymerise to form starch are (d) ______________ molecules. The small molecule formed when starch is made is (e) ______________. Although PLA is a synthetic polymer, the lactic acid (f) ______________, is made by fermentation of sugar cane, a natural product. Polymers made in this way are often called (g) ______________.

Word bank

bioplastics, biopolymers, condensation, glucose, monomer, natural, water

Creative plastics

Some plastics have been discovered by accident, and others have been known for some time before they were put to use. What is certain is that chemists are continually looking to improve the properties of plastics in order that they can be used more and more in everyday life.

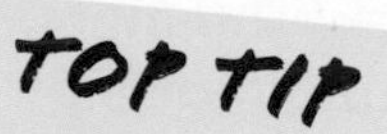

This section looks at modern developments in plastic chemistry and some unusual, but useful properties. You do not need to learn all the names and uses but be aware of some of the properties and how they can be used. The revision question at the end gives you practice in selecting and presenting information.

Kevlar is a plastic that was discovered years before a suitable solvent was found to dissolve it and make it usable. Kevlar polymer chains can be aligned to make **very strong** fibres that can be twisted to make cables, which have the same strength as steel cables but are five times **lighter**. Kevlar can be used in vehicle tyres instead of steel to improve the strength of the rubber and make the tyre lighter. Kevlar is well known as the protective padding in bullet-resistant vests. It is also used in some aircraft wings in which strength combined with being lightweight is essential.

Most of the kayaks used in the Olympics were made of Kevlar because it is tough and lightweight.

Polyvinyl alcohol (PVA) is a **soluble** plastic that has a huge number of uses. Solubility is not a property usually associated with plastics, but PVA combines solubility with strength, which makes it useful for making laundry bags for use in hospitals where infection may be a risk – the bag dissolves as the linen is washed and rinsed. Some detergents for use in dishwashers and washing machines are wrapped in PVA, which quickly dissolves releasing the detergent.

Sodium polyacrylate, often called **hydrogel**, is a polymer that **absorbs water**. Generally plastics repel water, which is one of their most useful properties. Chemists, however, saw the potential for using hydrogel beads in disposable nappies. Hydrogels can absorb 200–300 times their own weight of water. Chemicals that change colour when they are wet are sometimes added so that it can easily be seen if the nappy is leaking or very wet.

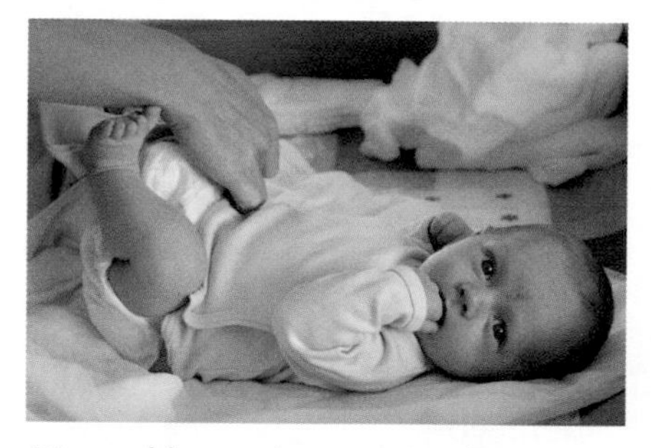

Disposable nappies contain plastic gel that absorbs water.

A group of scientists spent 17 years developing a plastic marketed as **RhinoPlex**, which is used to instantly repair tyre punctures. The plastic is sprayed onto the inside of the tyre and if the tyre is punctured the plastic immediately fills the hole, stopping air from leaking out.

'Smart' plastics that react to their environment are being developed. Chemists in the USA have developed a **self-healing plastic** that mimics the skin's ability to heal scratches and cuts. It could potentially be used as a self-repairing surface for the likes of mobile phones and laptops. The plastic **changes colour**, turning red to warn that there is damage to the surface, and will then heal itself when exposed to light. These smart plastics have great potential for use in areas where safety is a major issue, such as in aircraft wings and bridges. If a crack developed it would turn red so it could be easily spotted. Engineers could then investigate and decide whether to repair or replace the damaged area.

The incorporation of **colour-changing plastics** into food packaging materials could be used to let consumers know the conditions inside a food package. Plastics that are extremely sensitive to **stress** have been developed. The transparent plastic is stretched across the food, and if the food starts to go off, gases are given off, causing bloating of the plastic film. This causes the molecules in the plastic film to stretch out, and they interact with light in such a way as to cause colour to appear in the plastic. The appearance of colour would let the customer know that there could be a problem with the food. This colour-changing technology could replace coloured dyes as indicators of food quality.

Heat-sensitive plastics that change colour as temperature changes have been developed.

Plastics usually conduct electricity so poorly that they are used to insulate electric cables. However chemists have developed polymers that can be made to conduct electricity. Australian researchers have shown that by placing a thin film of metal into a plastic sheet and mixing the metal into the plastic using an ion beam they can make cheap, strong, flexible and **conductive plastic** films. It is predicted that this discovery has the ability to open up new avenues for making **plastronics** – plastic electronics. Conductive plastics could see flexible touch screens and e-paper a realistic possibility in the near future.

Quick Test

1. Summarise the information about creative plastics in a table. Make sure you include the name or type of polymer, important properties and how the property is used to our advantage.

Ammonia – the Haber process

Remember from National 4!

Plants need nutrients for healthy growth. Nitrogen is one of the essential elements for plant growth.

Making ammonia in the laboratory

TOP TIP

Ammonia turns pH indicator purple showing it is **alkaline**.

Ammonia (NH_3) is the most important source of nitrogen, both on its own and in compounds.

Ammonia is a **gas** and can be made in the laboratory by heating an ammonium salt, such as ammonium chloride, with a base like solid calcium hydroxide or sodium hydroxide solution:

$$2NH_4Cl(s) + Ca(OH)_2(s) \rightarrow CaCl_2(s) + 2H_2O(l) + \mathbf{2NH_3(g)}$$

This way of making ammonia was one of the methods used before the Haber process was invented.

Ammonia is extremely **soluble** in water. This can be demonstrated in the fountain experiment. A dry flask filled with ammonia gas is inverted in a beaker of water.

Ammonia has a very strong smell, which can affect your breathing – it is the smell often associated with wet nappies.

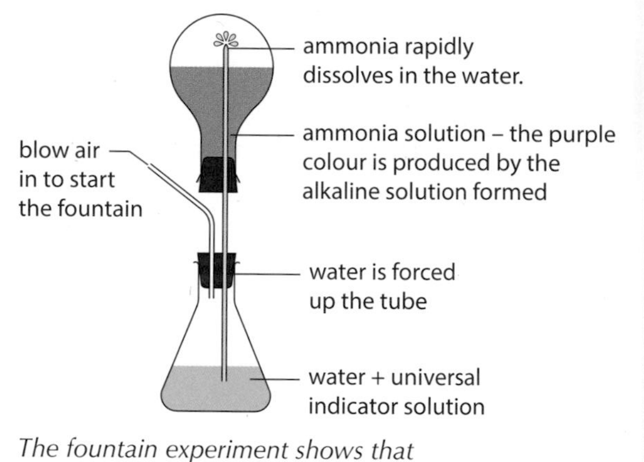

The fountain experiment shows that ammonia is very soluble.

Industrial Production: the Haber process

Industrially, ammonia is made by the direct combination of nitrogen from the air and hydrogen obtained from natural gas.

$$N_2(g) + 3H_2(g) \rightleftharpoons 2NH_3(g)$$

nitrogen + hydrogen ⇌ ammonia

This is known as the Haber process, named after its inventor Fritz Haber.

It is well known that changing the **temperature** affects the rate of reactions, but if too high a temperature is used it causes the ammonia to break down. However, too low a temperature slows the rate at which the ammonia is formed.

Because the reactants and products are gases, increasing the **pressure** produces more ammonia.

The diagram summarises the Haber process:

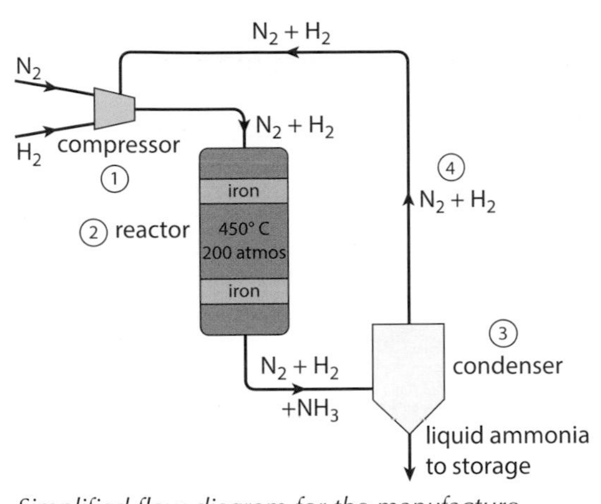

Simplified flow diagram for the manufacture of ammonia.

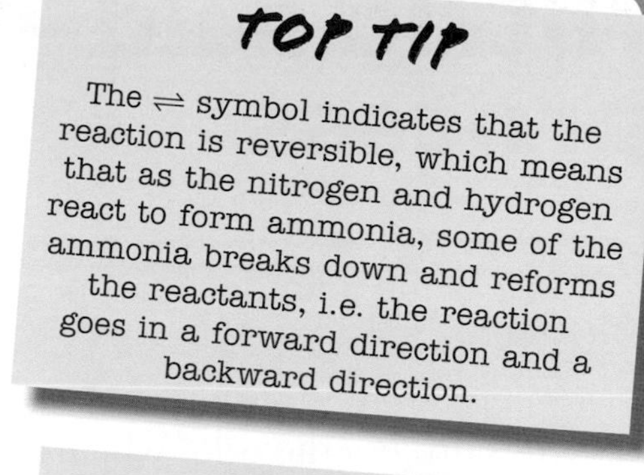

TOP TIP

Using a finely divided catalyst means there is a greater surface area for the reaction to take place on. The temperature is moderately high to make sure the ammonia is produced at an economical rate.

1. Nitrogen (from the air) and hydrogen (from natural gas) are compressed to a pressure of around **200 atmospheres** – 200 times higher than normal air pressure.
2. The compressed gases flow into the main reactor where they pass over a finely divided **iron catalyst** at about **450°C**.
3. The ammonia is passed into a condenser where it is cooled so that it **liquefies**.
4. Unreacted nitrogen and hydrogen are **recycled**. This saves valuable chemicals and helps push the yield of ammonia up to an economical amount (around 25%).

Quick Test

1. (a) Describe how ammonia gas could be made from ammonium sulfate.

(b) How could you test the gas to show it was ammonia?

(c) Suggest why any piece of equipment used to collect ammonia has to be dry.

2. Ammonia is produced industrially from its elements by the Haber process.

(a) Write a balanced equation for the production of ammonia from its elements.

(b) Summarise the conditions used in the Haber process.

(c) Explain why there is such a low conversion rate of nitrogen and hydrogen to ammonia.

(d) Explain why the temperature used in the process is not made lower.

(e) Unreacted nitrogen and hydrogen are recycled. Why is this an important part of the process?

(f) Explain why the catalyst is finely divided rather than being a solid piece of metal.

Nitric acid – the Ostwald process

Some of the **ammonia** produced by the Haber process is used to make **nitric acid (HNO_3)**.

It might seem more obvious to try to combine nitrogen and oxygen from the air to form nitrogen dioxide, which then dissolves in water to give nitric acid. Although it seems straightforward, the problem of nitrogen's lack of reactivity means that huge amounts of electricity are needed to supply the energy required to make the nitrogen react.

Wilhelm Ostwald invented the industrial method of converting ammonia into nitric acid, known as the **Ostwald process**.

He found that if ammonia and air were passed over a heated catalyst, nitrogen dioxide was formed, which dissolved in water to form nitric acid.

The diagram gives a simplified outline of the Ostwald process.

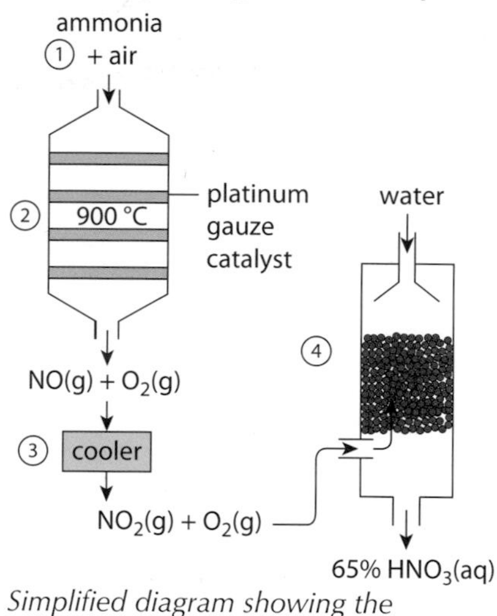

Simplified diagram showing the manufacture of nitric acid.

Stage 1: A mixture of **ammonia** and **air** is compressed up to 14 atmospheres.

Stage 2: The ammonia/air mixture is passed through layers of **platinum gauze catalyst** at about **900°C** and **nitrogen oxide (NO)** is formed. The gauze provides a large surface area for the reaction to take place and allows the gases to move freely through it.

Stage 3: The nitrogen oxide and air are passed through a cooler where they react to form the brown gas **nitrogen dioxide (NO_2)**.

Stage 4: The nitrogen dioxide and air are passed up an absorption tower packed with small beads. Water falls freely over the beads and mixes with the nitrogen dioxide and air. Dilute **nitric acid (HNO_3)** is formed.

The equations for the reactions happening in stages 2–4 are:

Stage 2:

ammonia + oxygen → nitrogen monoxide + water

$4NH_3(g) + 5O_2(g) \rightarrow 4NO(g) + 6H_2O(g)$

Stage 3:

nitrogen monoxide + oxygen → nitrogen dioxide

$2NO(g) + O_2(g) \rightarrow 2NO_2(g)$

Stage 4:

nitrogen dioxide + oxygen + water → nitric acid

$4NO_2(g) + O_2(g) + 2H_2O(l) \rightarrow 4HNO_3(aq)$

TOP TIP

The reaction between ammonia and oxygen in stage 2 is **exothermic**, so once the reaction has started, the external heating source can be reduced because the reaction produces its own energy. This saves a lot of energy so reduces costs.

Quick Test

1. The diagram shows a method for carrying out the Ostwald process in the laboratory.

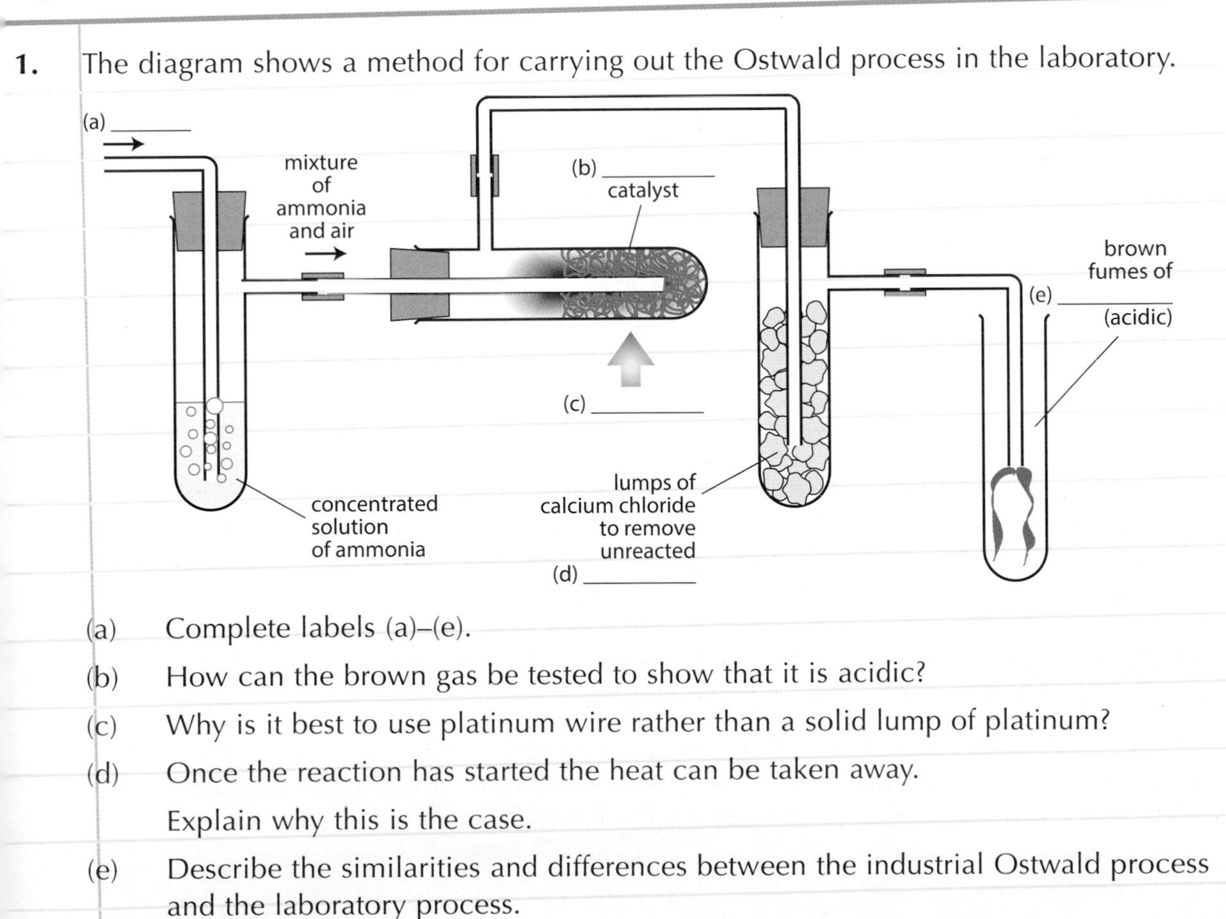

(a) Complete labels (a)–(e).

(b) How can the brown gas be tested to show that it is acidic?

(c) Why is it best to use platinum wire rather than a solid lump of platinum?

(d) Once the reaction has started the heat can be taken away.
Explain why this is the case.

(e) Describe the similarities and differences between the industrial Ostwald process and the laboratory process.

Synthetic fertilisers

Remember from National 4!

Plants need nutrients in the form of the essential elements nitrogen (N), phosphorus (P) and potassium (K) for healthy growth. There are not enough natural fertilisers like compost and manure to supply the amount of nutrients needed to meet world demand for food. Synthetic fertilisers containing various percentages of N, P and K are made by chemists to help meet demand.

Ammonium nitrate

Most of the ammonia and nitric acid produced in the world is used to make **ammonium nitrate**, which is one of the most widely used **synthetic fertilisers** in Europe.

Ammonium nitrate is made by reacting ammonia with nitric acid in a **neutralisation** reaction:

ammonia + nitric acid → ammonium nitrate

$$NH_3(g) + HNO_3(aq) \rightarrow NH_4NO_3(aq)$$

The industrial production of ammonium nitrate is often carried out on the same site as ammonia and nitric acid manufacture. This makes sense, as the ammonia is used to make nitric acid and ammonium nitrate. The flow diagram shows the simplified production process of ammonium nitrate.

A good synthetic fertiliser has to contain the essential elements **N, P** and **K,** and be **soluble** in water so that the nutrients can be taken into the plant through its root system.

Ammonium nitrate's high solubility means it is easily washed out of the soil, which can lead to pollution problems.

NPK mixtures are commonly used by farmers.

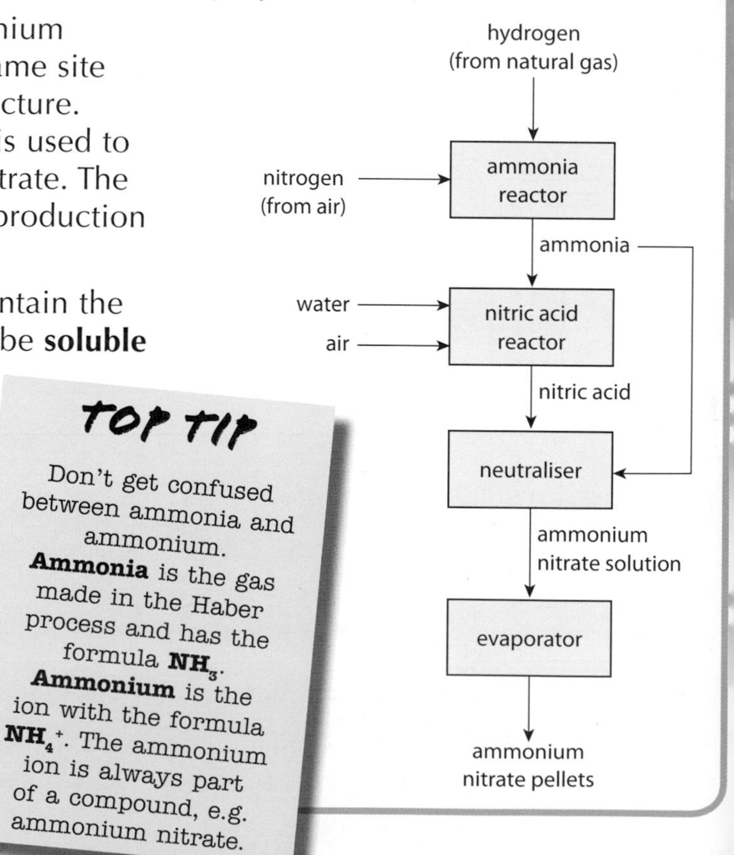

TOP TIP

Don't get confused between ammonia and ammonium. **Ammonia** is the gas made in the Haber process and has the formula NH_3. **Ammonium** is the ion with the formula NH_4^+. The ammonium ion is always part of a compound, e.g. ammonium nitrate.

Calculating the percentage mass composition of elements in fertilisers

The table gives examples of common synthetic fertilisers and their percentage composition of the essential elements by mass.

Fertiliser	Formula	%N	%P	%K
ammonium phosphate	$(NH_4)_3PO_4$	28	21	0
potassium nitrate	KNO_3	39	0	14
ammonium sulfate	$(NH_4)_2SO_4$	21	0	0
urea	$(NH_2)_2CO$	47	0	0

Percentage composition by mass of the essential elements in some common fertilisers

The percentage composition by mass of the essential elements N, P and K found in common fertilisers can be calculated from the relative atomic mass (RAM) of the elements in the compound as shown for ammonium nitrate:

% composition by mass of N = (mass of N / formula mass of NH_4NO_3) × 100

$$= \frac{(2 \times 14)}{(2 \times 14) + (4 \times 1) + (3 \times 16)} \times 100$$

$$= \frac{28}{80} \times 100$$

% composition by mass of N in ammonium nitrate = 35%

Quick Test

1. Look at the flow diagram for the manufacture of ammonium nitrate.
   (a) Write a word and formula equation for the reaction happening in the neutraliser.
   (b) Describe what happens in the evaporator.
2. Ammonium phosphate is used as a synthetic fertiliser.
   (a) Write word and formulae equations for the formation of ammonium phosphate from ammonia and phosphoric acid (H_3PO_4).
   (b) What type of reaction is taking place in (a)?
   (c) State two things about ammonium phosphate that make it a good fertiliser.
3. Potassium sulfate (K_2SO_4) could be used as a fertiliser. Calculate the percentage composition by mass of potassium in potassium sulfate.

Radioactivity and radioisotopes

The atoms of most elements have **isotopes**. The nuclei of some of these isotopes are unstable and give out particles and rays, called **emissions**. This is known as **radioactivity**. Radioactivity happens spontaneously no matter what state the element is in or if it is chemically combined in a compound. These emissions are called **alpha** (α), **beta** (β) and **gamma** (γ).

The diagrams show the **penetrating power** of the radioactive emissions and what happens to them when they are passed through an **electric field**.

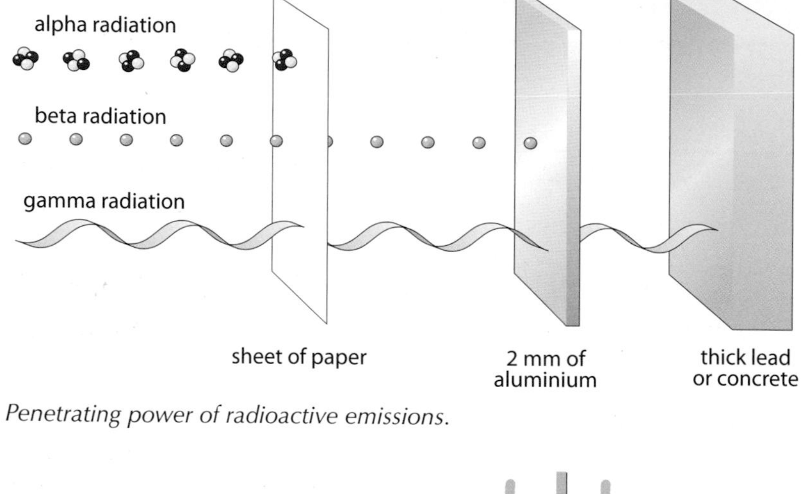

Penetrating power of radioactive emissions.

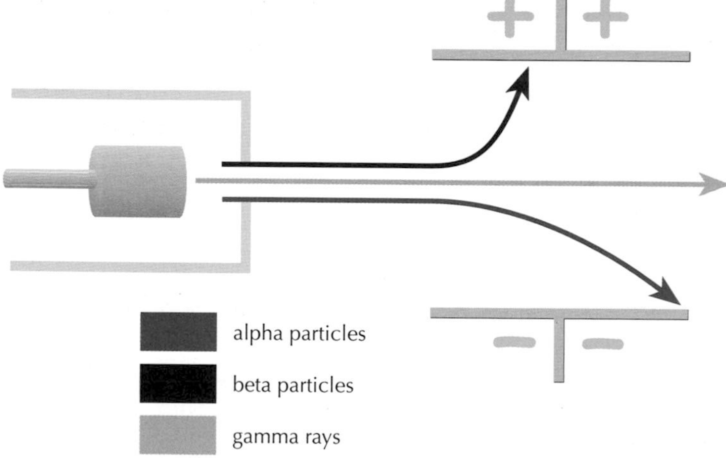

Deflection of emissions in an electric field.

The properties of radioactive emissions are summarised in the table.

Name of emission	Distance travelled and penetrating power	Charge
Alpha (α)	Few centimetres in air. Stopped by paper.	positive
Beta (β)	Few metres in air. Stopped by thin aluminium sheet.	negative
Gamma (γ)	Miles in the air. Stopped by thick lead or concrete.	none

Properties of radioactive emissions

α **particles** are the heaviest of the emissions and are made up of two protons and two neutrons grouped together. An α particle is in fact a helium ion – He^{2+}. Its nuclide notation is $^{4}_{2}\mathbf{He}^{2+}$ – the charge is often not shown.

β **particles** are high-energy electrons. A β particle is often represented as $^{0}_{-1}\mathbf{e}$.

γ **radiation** is not a particle and so has no mass and no charge. It is a **high-energy ray** that can travel long distances and has high penetration power.

Using radioactive isotopes

Radioactive isotopes are also known as **radioisotopes** and they are important in many everyday situations.

In the **home**: radioisotopes have limited use in the home but play a vital role in smoke alarms. They contain americium-241, which emits α particles that cause an electric current to flow. When smoke passes into the alarm's detector, the current drops and the alarm sounds.

In the **health service**: radioisotopes are used in a variety of ways. Cancer in the thyroid gland in the neck can be **detected** and **treated** using isotopes of iodine, which emit γ and β radiation.

In **industry**: radioisotopes are widely used to control the thickness of materials – this is known as **gauging**. The material passes between a radioactive source and a detector. The intensity of the radioactivity reduces when it hits the material. This is picked up by the detector, and the thickness of the material can be adjusted as necessary.

The fear of radioactivity damaging healthy cells and causing cancer is one of the main reasons for many people being concerned about the use of radioisotopes.

Quick Test

1. Complete the summary by filling in the missing words. You can use the word bank to help you.

Atoms with unstable (a)__________ can emit (b)__________ (α), beta (β) and (c)__________(γ) radiation. α particles are (d)__________ nuclei and β particles are high-energy (e)__________. α particles are (f)__________ moving and can be stopped by paper. (g)__________ particles are fast moving and can be stopped by a thin sheet of (h)__________ γ rays can travel long distances and can only be stopped by thick lead or concrete. Radioisotopes have many everyday uses. They are used in most (i)______________ alarms in the home. In industry, (j)__________ the thickness of materials is controlled by radioisotopes. In the health service they are used in the detection and treatment of (k)__________. Radiation can (l)__________ healthy cells in the body as it passes through them.

Word bank

alpha, aluminium, beta, cancer, damage, electrons, gamma, gauging, helium, ionises, nuclei, slow, smoke

Nuclear equations and half-life

Nuclear equations

The breakdown of unstable nuclei (radiation) is also known as **radioactive decay**. An atom can decay through a series of stages until it forms atoms of a stable isotope.

The emission of an α particle means the nucleus of the original atom loses two protons and two neutrons. The emission of a β particle results in the original atom gaining a proton. γ radiation has no effect on the original atom because it is not a particle.

Radioactive decay can be represented by **nuclear equations** – they can be used to summarise the processes that produce α and β radiation.

Nuclear equations include the **mass number** (number of protons + neutrons), the **atomic number** (number of protons) and the **chemical symbol** for each particle involved, i.e. the nuclide notation.

Example of α **decay:**

When plutonium-242 decays by α ($^{4}_{2}He$) emission, uranium-238 is formed.

Nuclear equation: $^{242}_{94}Pu \rightarrow {}^{238}_{92}U + {}^{4}_{2}He$

Example of β **decay:**

When thorium-234 decays by β ($^{0}_{-1}e$) emission, protactinium-234 is formed.

Nuclear equation: $^{234}_{90}Th \rightarrow {}^{234}_{91}Pa + {}^{0}_{-1}e$

TOP TIP

The total of the mass number on the left of the arrow must equal the total on the right of the arrow. It is the same for the atomic numbers. Atomic numbers are in the SQA data booklet.

Note again that when the mass numbers and atomic numbers are added on each side of the arrow they are the same. The electron (β particle) is given an unusual atomic number (–1). This is a way of indicating that an extra proton is gained by the parent atom when a beta particle is emitted and so the rule that the total atomic number must be the same on each side of the arrow is satisfied.

So long as you know the particle being emitted, the element formed can be worked out, as illustrated in the following examples.

Example 1: Work out what a, b and X are in the nuclear equation:

$^{220}_{86}Rn \rightarrow {}^{a}_{b}X + {}^{4}_{2}He$

Worked answer: Apply the rule that mass and atomic numbers must add up to the same on each side of the equation.

So, a = 216; b = 84 so X must be Po (polonium), i.e. $^{216}_{84}Po$

Example 2: Work out what c, d and Z are in the nuclear equation:

$^{228}_{88}Ra \rightarrow {}^{c}_{d}Z + {}^{0}_{-1}e$

Worked answer: Apply the rule that mass and atomic numbers must add up to the same on each side of the equation.

So, c = 228; d = 89 so Z must be Ac (actinium), i.e. $^{228}_{89}Ac$

Artificial radioisotopes can be made in nuclear reactors by bombarding stable nuclei with neutrons. For example:

$^{27}_{13}Al + ^{1}_{0}n \rightarrow ^{24}_{11}Na + ^{4}_{2}He$

The sodium isotope produced is radioactive and decays itself by β emission:

$^{24}_{11}Na \rightarrow ^{24}_{12}Mg + ^{0}_{-1}e$

Radioactive phosphorus-32 is produced by neutron bombardment of sulfur-32. A proton ($^{1}_{1}p$) is also produced:

$^{32}_{16}S + ^{1}_{0}n \rightarrow ^{32}_{15}P + ^{1}_{1}p$

Many artificial radioisotopes are produced for specific uses in health care and industry.

TOP TIP

Learn the notation for particles: neutron: $^{1}_{0}n$; proton: $^{1}_{1}p$; β (electron): $^{0}_{-1}e$; α: $^{4}_{2}He$.

Half-life

The nuclei of radioisotopes decay in a random fashion. The time in which half of the nuclei of a radioisotope would be expected to decay is known as the **half-life,** often abbreviated to $\mathbf{t_{1/2}}$. Each radioisotope has a unique half-life and this can vary from fractions of a second to millions of years.

Radioisotope	Half-life	Use
Iodine-131	8.02 days	Diagnosing/treating diseases associated with the thyroid gland
Phosphorus-32	14.28 days	Treatment of excess red blood cells
Technetium-99m	6.01 hours	Imaging the organs of the body
Americium-241	433 years	Smoke detectors
Cobalt-60	5.27 years	Gamma radiography and medical-equipment sterilisation
Caesium-137	30.07 years	Thickness gauging

Selection of radioiotope half-lives

As the atoms of a radioisotope decay, the intensity of the radiation decreases. After one half-life the intensity of the radiation will have fallen to half its original value. After a second half-life the intensity of the radiation will have halved again, i.e. it will be one quarter of its original value. A graph of the intensity of the radiation emitted against time gives a radioactive **decay curve** with a typical shape.

The graph shows the change in activity for a radioisotope where $t_{1/2}$ = two days.

The graph shows it takes two days (one half-life) for the intensity of the radiation to halve. After two half-lives (four days) the intensity of the radiation will have dropped to a quarter of its original value and so on.

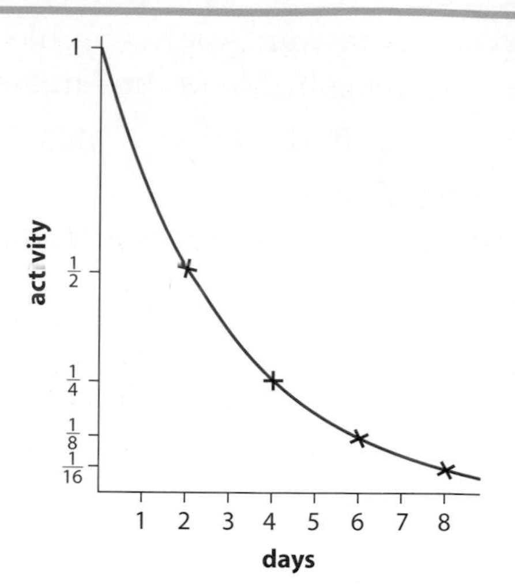

Decay curve for a radioisotope with a half-life of two days.

So, if you start with 1 g of the sample, after one half-life (two days) half of the original sample (0.5 g) will have decayed, leaving the other 0.5 g active. After another half-life (four days total) half of the remaining 0.5 g will decay, leaving 0.25 g of the original sample. After a third half-life (six days) half of the 0.25 g will have decayed, leaving 0.125 g of the original sample. So, after three half-lives only 0.125 g of the original sample will be left.

Worked example: Sodium-24 has a half-life of 15 hours. What mass of a 0.4 g sample of sodium-24 would be left after 60 hours?

Answer: Number of half-lives = 60/15 = $4t_{1/2}$

After $1t_{1/2}$ (15 hours), 0.2 g left

After $2t_{1/2}$ (30 hours), 0.1 g left

After $3t_{1/2}$ (45 hours), 0.05 g left

After $4t_{1/2}$ (60 hours), 0.025 g left.

Dating using radioisotopes

To work out the age of material, researchers compare the ratio of a radioactive isotope of an element present in a sample with the proportion of stable isotope present and compare the ratio with a sample of the same material at the present time. By doing this and knowing the half-life of the radioisotope they can calculate how much time has passed.

Radiocarbon dating

Carbon dating is one of the most widely known uses of radioisotopes. It is used in archaeological research and can be used to date specimens containing material that was once living. All living things contain a small amount of unstable carbon-14 (^{14}C) and stable carbon-12 (^{12}C).

Atoms of ^{14}C are radioisotopes and are formed in the upper atmosphere when neutrons in cosmic rays collide with nitrogen atoms.

$$^{14}_{7}N + ^{1}_{0}n \rightarrow ^{14}_{6}C + ^{1}_{1}p$$

The carbon atoms react with oxygen in the air to form carbon dioxide, which is absorbed by plants during photosynthesis, and are therefore passed into food chains. All living things therefore contain a small amount of unstable ^{14}C and stable ^{12}C. The ratio of ^{14}C:^{12}C in a living organism remains constant throughout its lifetime as it is continually taking in carbon dioxide. When an organism dies, the amount of ^{14}C in the organism begins to decrease as ^{14}C atoms decay but the stable ^{12}C does not. ^{14}C decays by beta emission and has a half-life of approximately 5730 years.

$$^{14}_{6}C \rightarrow ^{14}_{7}N + ^{0}_{-1}e$$

By measuring the ^{14}C:^{12}C ratio in the material and comparing it to the current ratio in living material, the time since the organism died can be measured.

Radiocarbon dating is thought to be accurate up to 50 000 years. The technique has been particularly useful in dating ancient Egyptian relics and important historical items such as the Dead Sea Scrolls and the Turin Shroud.

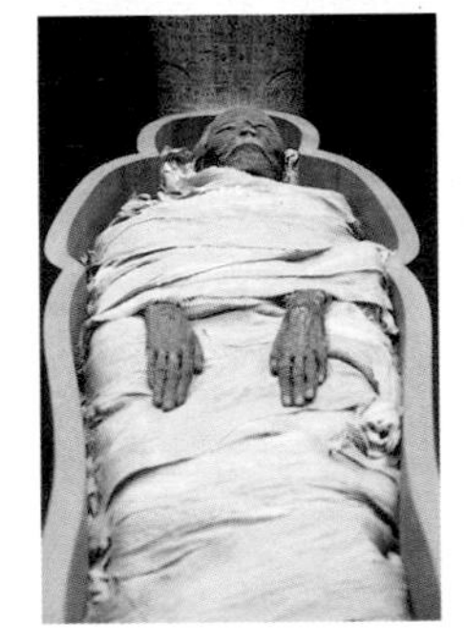

Dating rocks and fossils

Radiocarbon dating is only useful for materials less than about 50 000 years old owing to the very small amount of ^{14}C present in materials older than this.

Other methods making use of radioisotopes with much longer half-lives have to be used to date rocks and fossils.

One system that has been very successful in dating the ages of fossils is **potassium–argon dating**. Potassium is an extremely common element. Although most potassium isotopes aren't radioactive, one of them is, and one of its decay products is the gas argon.

$$^{40}_{19}K + ^{0}_{-1}e \rightarrow ^{40}_{18}Ar + \gamma$$

The half-life for this transition is 1.3 billion years.

Although potassium is a solid, argon is a gas. When rock is melted (lava), all the argon in the rock escapes, and when the rock solidifies again only potassium is left. As time passes, argon forms in the rock as a result of radioactive potassium decay. When scientists analyse these rocks and work out the ratio of argon to potassium, they can calculate how long it has been since the lava cooled. By dating the lava flows above and below a fossil find, scientists can accurately estimate the age of the fossil. Dinosaur remains from millions of years ago found trapped between layers of rock have been dated this way.

Dating the Earth

Scientists are unable to deduce the age of the Earth from rocks because Earth rocks have been formed and re-formed over time. They have worked out the age of the Earth by dating meteorites they believe were formed at the same time as the Earth. Isotopes of uranium in meteorites decay to form isotopes of lead. Measuring the proportions of lead isotopes to uranium isotopes gives an indication of the age of the meteorites.

The largest meteorite ever found weighed 60 tonnes and was found in Namibia but larger meteorites have struck the Earth. The Canyon Diablo meteorite, which caused the formation of a meteor crater in Arizona when it struck the Earth, is estimated to have weighed over 60 000 tonnes. Analysis of the meteorite indicates that the Earth is about 4.5 billion years old.

Moon rock and samples from Mars have been analysed and their ages estimated to be around the same as the Earth, supporting the idea that our solar system was formed at the same time.

Quick Test

1. Write nuclear equations for the following:
   (a) The α decay of $^{234}_{92}U$.
   (b) The β decay of $^{228}_{89}Ac$.
   (c) The decay of ^{210}Bi to ^{210}Po.
   (d) The neutron bombardment of $^{45}_{21}Sc$, which produces $^{42}_{19}K$ and an α particle.
2. The decay curve for strontium-90 is shown:

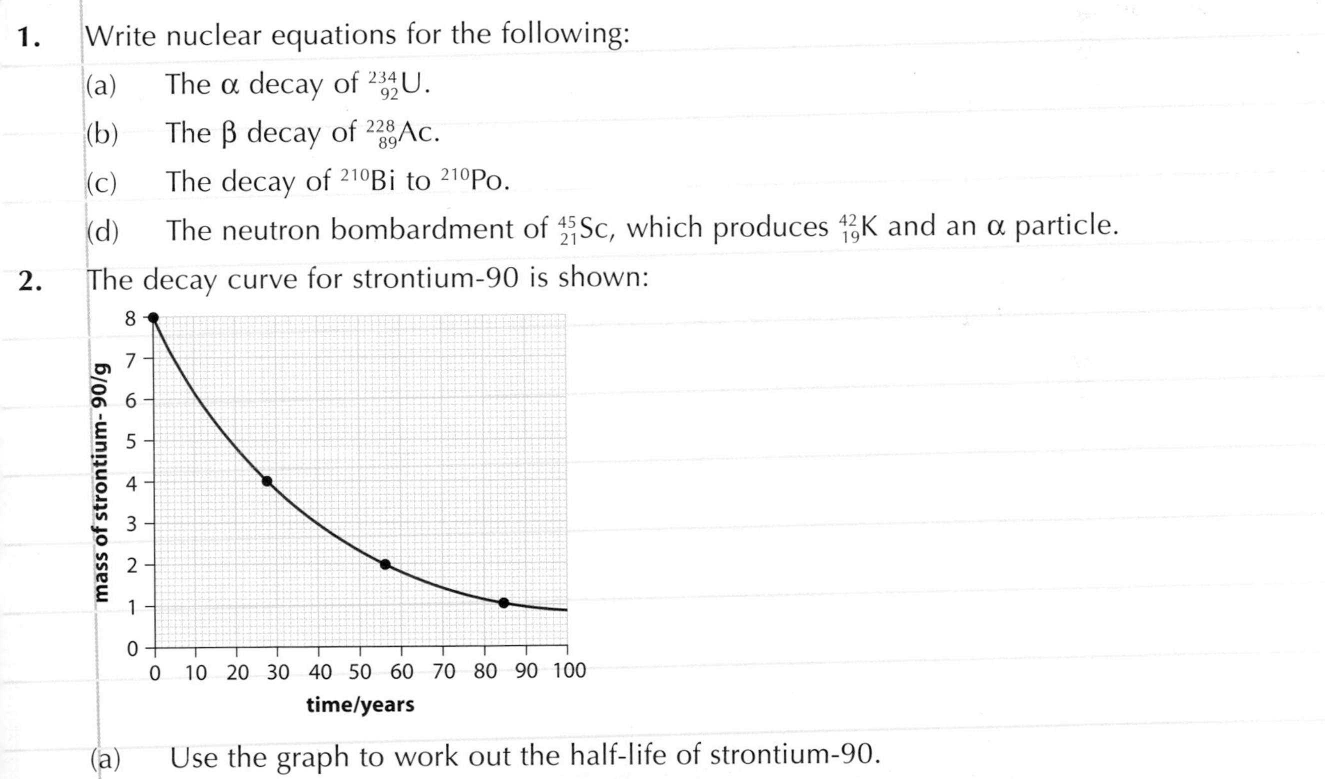

   (a) Use the graph to work out the half-life of strontium-90.
   (b) How long would it take for the mass of the strontium-90 to fall to 2.5 g?
   (c) What mass of strontium-90 would be left after $t_{1/2} = 3$?
3. The half-life of carbon-14 is 5730 years.
   (a) How long will it take for the radioactive count rate to drop to 25% of its original value?
   (b) Outline how carbon-14 can be used to date a piece of wood found on an archaeological dig.

Chemical analysis

The role of chemists

The UK horse-meat scandal in early 2013 emphasised the importance of monitoring our food industry. Chemists have an important role in monitoring our environment, too. They monitor **air** and **water** quality as well as **soil** conditions to ensure that farmers can grow crops under the best conditions. Pollution levels in the environment need to be monitored because of the potential impacts on human health and on the natural environment itself.

Pollution from fertiliser run-off from fields has caused **algal blooms** in rivers and lochs leading to the death of aquatic plants and fish. Loss of fertilisers from the fields is also uneconomical and results in poor crop growth.

Air pollution from domestic and industrial sources is tending to decrease. However air pollution caused by transport is increasing worldwide. The **main air pollution** problem nowadays comes from the **increasing number of vehicles** on our roads. Burning petrol, diesel and kerosene (jet fuel) emits a variety of pollutants into the atmosphere. These include carbon monoxide (CO), oxides of nitrogen (NO_x), unburnt hydrocarbons and tiny solid particles, which can pass through the nose and throat and enter the lungs. This is a particular problem in towns and cities.

The concentrations of a number of air pollutants are continuously and automatically monitored.

Nitrogen dioxide levels can be measured by absorbing the gas in NO_2-collection tubes. When the monitoring period is over, the chemical in the tube is reacted with other chemicals to give a coloured solution. The intensity of the colour can be measured using a spectrometer or colorimeter giving a value for the nitrogen dioxide level.

Satellites orbiting above the Earth are also used to detect pollutants in the air.

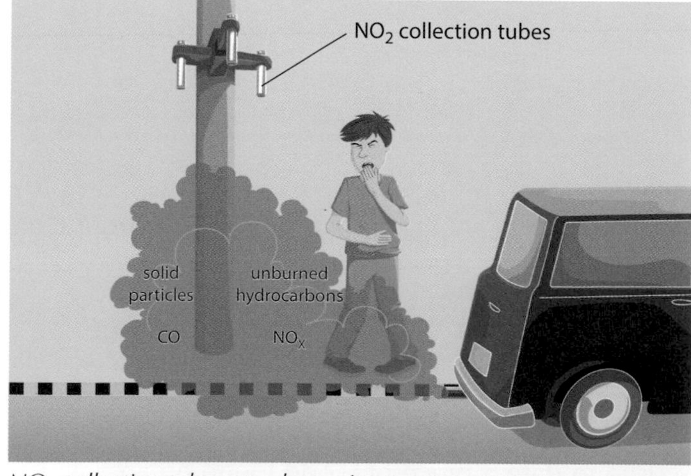

NO_2-collection tubes on a busy city street.

Dealing with pollution problems as they arise

Chemists and chemical engineers have an important role in thinking up ways of **removing harmful chemicals** and thus preventing them affecting the environment.

Many industrial plants such as coal-burning power stations remove **sulfur dioxide** by a process known as **flue-gas desulfurisation** (FGD). Flue gases are passed through a **scrubber tower** before being released into the atmosphere.

Sulfur dioxide is an **acidic** gas and is very soluble. If flue gases are passed through an **alkaline** spray the sulfur dioxide will react. Some wet scrubbers use a limestone slurry (finely powdered limestone in water). The sulfur dioxide is converted to calcium sulfite.

$$CaCO_3(s) + SO_2(g) \rightarrow CaSO_3(s) + CO_2(g)$$

The calcium sulfite can be oxidised to calcium sulfate by blowing compressed air through it. Some of the costs of FGD can be recouped by selling the calcium sulfate, which is used to make plasterboard for the building trade.

Carbon dioxide can also be reacted with an alkali to remove it. Increasing levels of carbon dioxide are contributing to global warming so can't be released into the atmosphere.

Catalysts that absorb light energy, known as **photocatalysts**, help break down air pollutants into harmless chemicals. Photocatalysts have already been incorporated into coatings for concrete and self-cleaning glass in new buildings. It has also been shown that clothing, like denim jeans, with photocatalysts on their surface could also break down air pollutants as you walk in the street.

Quick Test

1. Explain why it is important that chemists monitor what is happening to the environment.
2. Air pollution is on the increase in some parts of the world.
   (a) State the main cause of the increase in air pollution and name the main pollutants.
   (b) How is air quality being monitored across the country?
   (c) Outline one way in which sulfur dioxide produced industrially can be prevented from reaching the atmosphere.
   (d) Give an example of how pollutants that get into the air are being broken down using catalysts.

Qualitative analysis

Laboratory qualitative analysis techniques

Detecting substances that are present in our environment is known as **qualitative analysis**. The techniques used nowadays are fairly sophisticated but their origins can be traced back to simple laboratory experiments. Flame testing, precipitation reactions and chromatography are all examples of qualitative analysis techniques used in the laboratory.

Flame testing

When some metal compounds are placed in a flame, characteristic colours are observed. For instance when **sodium** compounds are placed in a flame a **yellow** colour is observed. **Potassium** gives a **lilac** flame. When the metal compounds are heated some electrons of the metal ions gain energy. The flame colours arise when the electrons lose the energy again. The energy is emitted as light of a certain colour, which is different for each metal.

TOP TIP

Flame colours can be found in the SQA data booklet.

A flame test.

Precipitation

Some **metal ions** can also be detected using **precipitation** reactions. When two solutions are mixed, and a solid forms, it is called a precipitate.

When **sodium hydroxide** solution is added to solutions containing metal ions **coloured** precipitates can be formed. If the solution contains iron(III) ions a rust-red precipitate of iron(III) hydroxide is formed.

$$3NaOH(aq) + FeCl_3(aq) \rightarrow \underset{\text{rust-red}}{Fe(OH)_3(s)} + 3NaCl(aq)$$

The colour of the precipitate helps identify the metal ion present.

Non-metal ions can also be detected using precipitation reactions.

When **silver nitrate** is added to solutions containing halide (group 7) ions, precipitates form. The **colour** of the precipitate indicates the particular halide ion present. When silver nitrate is added to a sodium **chloride** solution a white precipitate of silver chloride is obtained.

$$NaCl(aq) + AgNO_3(aq) \rightarrow NaNO_3(aq) + \underset{\text{white}}{AgCl(s)}$$

Precipitation reactions.

Chromatography

Chromatography can be used to separate and identify complicated mixtures of substances. There are many different chromatographic methods but they all work on the same principle.

In the lab the most common chromatography is simple paper chromatography. As the solvent moves up the filter paper it carries the substances with it. However the substances move up the paper at different rates depending on their attractions to the paper. The stronger these are, the slower the substance moves up. In paper chromatography the paper is known as the stationary phase, and the solvent moving up the paper is known as the mobile phase. All chromatography systems have a stationary phase and a mobile phase.

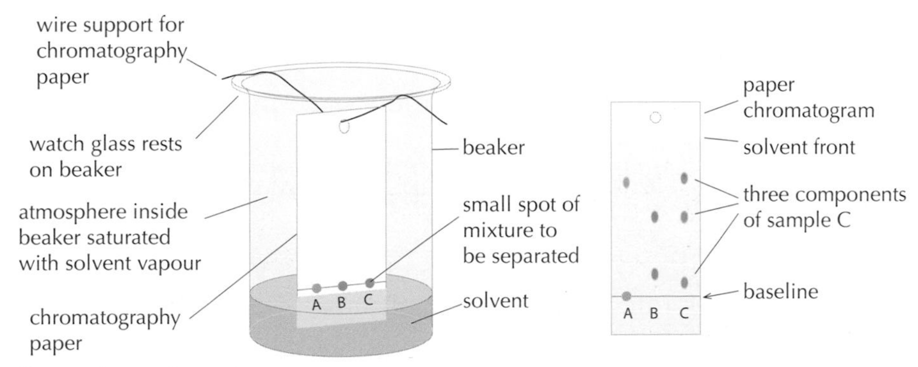

Paper chromatography.

Column chromatography was the first method to be used. A development of this method, called high-performance liquid chromatography (HPLC), has been used to detect additives and contaminants in food. When HPLC is linked to a mass spectrometer that can identify the separated components it yields a very powerful analytical tool. This has been used to detect steroids in athletes' bodies and contaminated food products.

Other chromatographic methods include gas–liquid, which is used to identify additives in soft drinks, and thin-layer chromatography, used widely in forensic science to identify the likes of explosives residues. Techniques are now so sensitive that they can detect minute quantities of drugs such as cocaine that someone may have come into contact with through touching contaminated material.

Genetic fingerprinting

DNA, the molecule that carries genetic information, can be detected using an analytical technique known as polymerase chain reaction (PCR). The technique is so accurate that it can be used to detect one molecule of DNA.

The chances of any two people having the same DNA are thought to be one in about 25 million. This forms the basis of genetic fingerprinting used by forensic scientists to identify criminals from blood and saliva. The DNA found at a crime scene is compared to that of a suspect, and if they match there is a very high probability that the suspect was at the crime scene.

PCR analysis was used in 2013 to prove that beef in some ready meals was contaminated with horse meat.

Space exploration

In 2012 the Curiosity Rover, launched by NASA, landed on Mars for a mission lasting two years to investigate whether Mars in the past was able to support life.

The instruments on board Curiosity are enabling scientists to find out more about the surface of the planet. The largest instrument, called Sample Analysis at Mars (SAM), is made up of three analytical instruments: a mass spectrometer, a gas chromatograph and a laser spectrometer.

SAM is being used to search for compounds of the element carbon, including methane, which is associated with life.

- The mass spectrometer separates elements and compounds by mass for identification and measurement.
- The gas chromatograph heats soil and rock samples until they vaporise, and then separates the resulting gases into various components for analysis.
- The laser spectrometer measures the abundance of isotopes of carbon, hydrogen and oxygen.

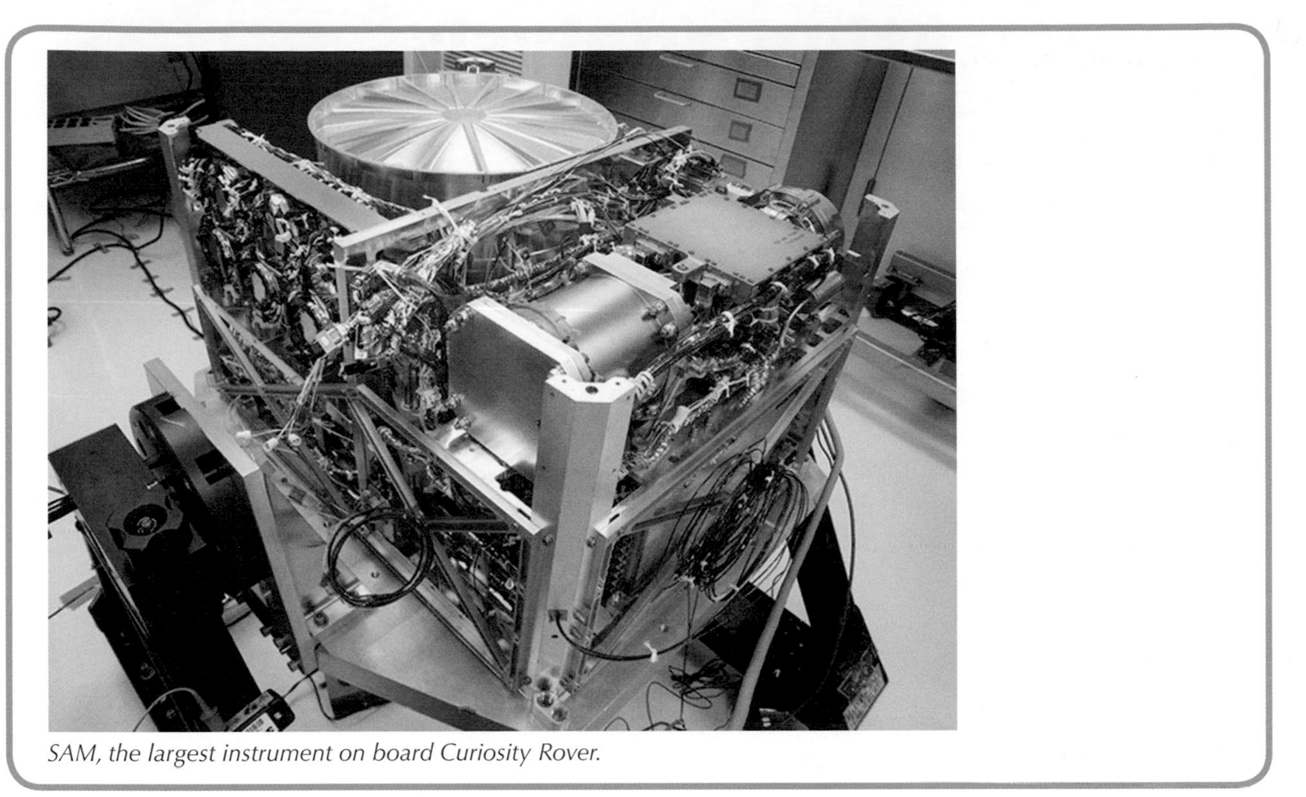

SAM, the largest instrument on board Curiosity Rover.

Quick Test

1. You are given soil samples from the Moon and asked to analyse them to find out if certain elements are present. Describe how you would test for:
   (a) potassium
   (b) iron(III) ions
   (c) halide ions.
2. A chemist uses paper chromatography to analyse a breakfast cereal to find out which sugars it contains.

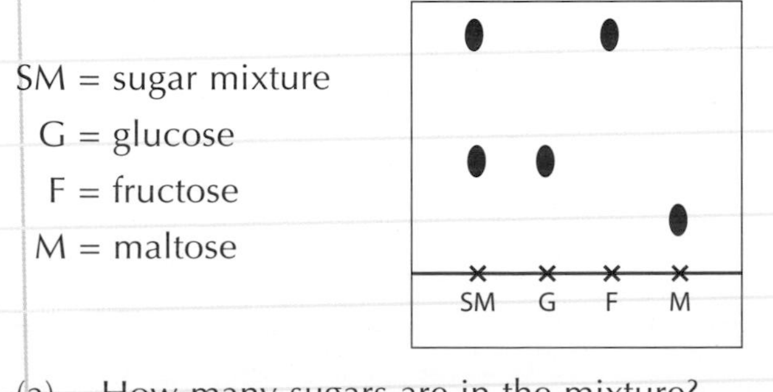

   (a) How many sugars are in the mixture?
   (b) Identify the sugars in the mixture.

Quantitative analysis

Remember from National 4!
pH measurement can be carried out using pH meters or indicators where different colours indicate different pH values.

Chemists are interested in which substances are present but also how much of these substances are present. This is **quantitative analysis**. One method used to work out how much of a substance is present in a solution is **titration**. Titration can be used to find the concentration of an acid or alkali using a neutralisation reaction.

How to do a titration

The concentration of a sodium hydroxide solution, say, can be found by titrating the solution with a dilute acid such as hydrochloric acid.

Step 1: A **pipette** is used to transfer a known quantity of the sodium hydroxide solution into a conical flask.

Step 2: Two or three drops of a suitable **indicator** are added to the solution and the flask placed on a white tile.

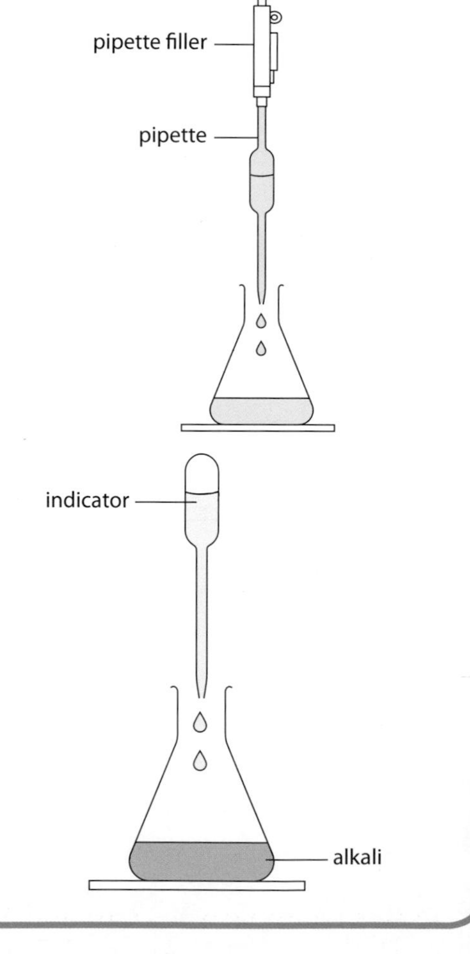

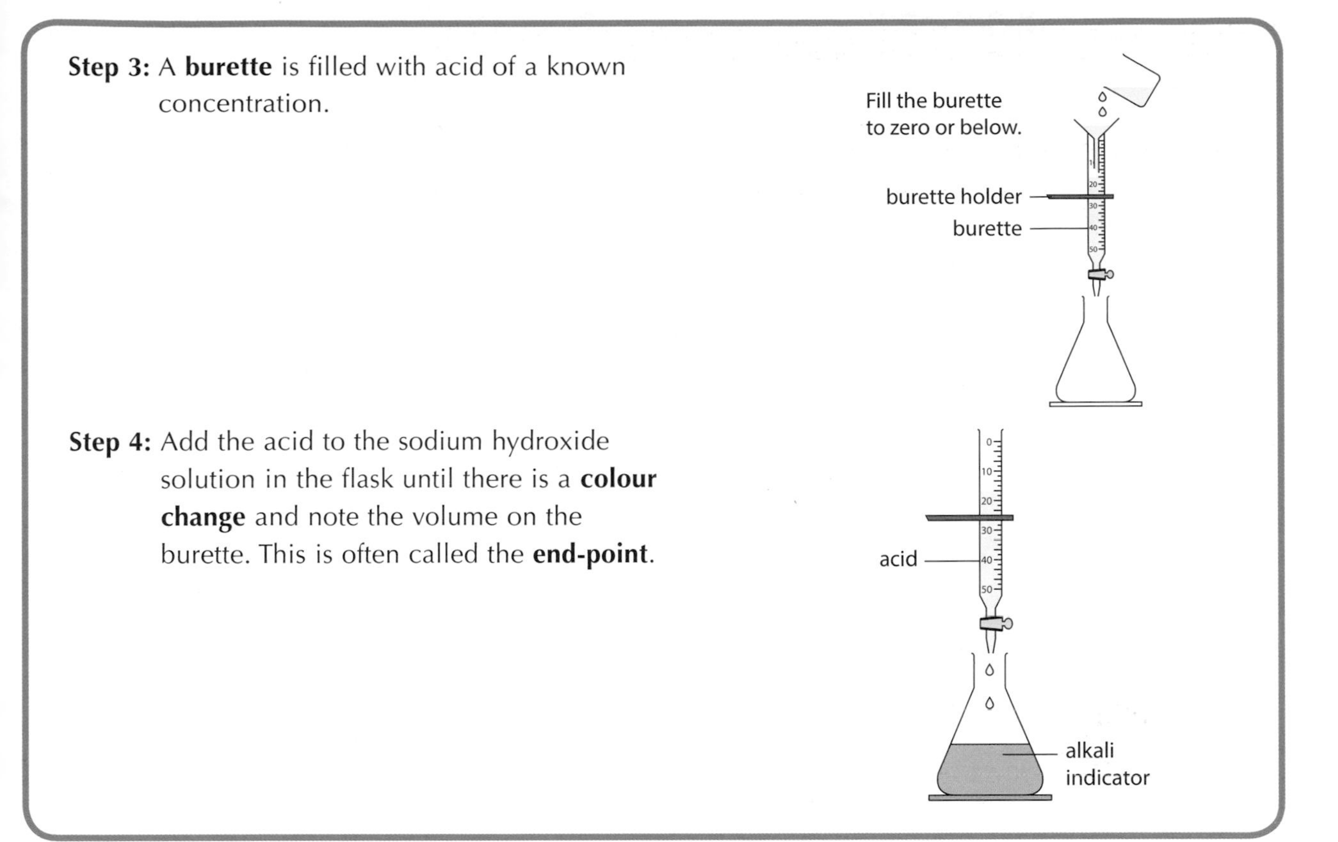

Accuracy counts

When titration is used in quantitative analysis, a rough titration to estimate the volume of solution that needs to be added is carried out. The titration is then carried out accurately to find the exact volume that needs to be added. It is good practice to repeat the titration until **concordant titres** are obtained. Concordant titres are where the volumes titrated are very close. The values are averaged (ignoring the rough titration) for use in the titration calculation.

Titration	1st level (cm^3)	2nd level (cm^3)	Volume added (cm^3)
1st (rough)	0.0	12.9	12.9
2nd	12.9	25.0	12.1
3rd	25.0	36.9	11.9

$$\text{Average} = \frac{12.1 + 11.9}{2}$$
$$= \underline{12.0\ cm^3}$$

In this example 12.1 cm^3 and 11.9 cm^3 are averaged because they are concordant values.

Titration calculation

Example: 12.0 cm^3 of 0.10 mol sulfuric acid was required to neutralise 20.0 cm^3 of potassium hydroxide solution. The balanced equation for the reaction is:

$2KOH(aq) + H_2SO_4(aq) \rightarrow K_2SO_4(aq) + 2H_2O(\ell)$

Calculate the concentration of the potassium hydroxide solution.

Worked answer: We can calculate the **number of moles** reacting then the concentration.

The balanced equation tells us that **1 mole** of sulfuric acid will neutralise **2 moles** of potassium hydroxide. We can use this and the relationship:

moles = concentration (moles per litre) × volume (litres)

to work out the unknown concentration.

Step 1: Work out the number of moles of sulfuric acid used in the titration.

Number of moles of acid = 0.1 mol 1^{-1} × 0.012 l (12 cm^3 = 0.012 l)

= 0.0012 mol.

Step 2: Work out the moles of potassium hydroxide that reacted.

This must be 0.0024 mol since **1 mol** of sulfuric acid reacts with **2 mol** of KOH.

Number of moles of potassium hydroxide = 0.0024 mol.

Step 3: Use the relationship:

concentration (mol 1^{-1}) = number of moles / volume (l)

= 0.0024/0.02 (20 cm^3 = 0.02 l)

= **0.12 mol l^{-1}**

Another way of setting this out is using the **balancing numbers** in the equation:

$$\mathbf{2}KOH(aq) + \mathbf{1}H_2SO_4(aq) \rightarrow K_2SO_4(aq) + 2H_2O(l)$$

The balancing number for potassium hydroxide is **2** and for sulfuric acid is **1**.

$$\frac{(\text{concentration (B)} \times \text{volume})_{\text{alkali}}}{\text{balancing no.}_{\text{alkali}}} = \frac{(\text{concentration} \times \text{volume})_{\text{acid}}}{\text{balancing no.}_{\text{acid}}}$$

$$\frac{B \times 20}{2} = \frac{0.1 \times 12}{1}$$

(Note: You don't have to change volumes to litres.)

B × 10 = 1.2

B = **0.12 mol l^{-1}**

TOP TIP

Don't try to learn both methods. Pick the one that suits you best and practise the method.

Using titration

Titration is used widely in industry. In wineries, for example, it is used to check the acidity of the wine, as this affects the keeping quality of the wine. In the dairy industry, titration is part of a procedure that measures the protein content of foods.

Titration allows the concentration of dissolved oxygen in water to be worked out. This is important in order to monitor the quality of water in lochs and rivers. Fish can't survive without good water quality. For trout and salmon the amount of oxygen in the water (known as dissolved oxygen) needs to be above 9 mg/l. If it falls below this level the fish will become stressed and there may not be enough oxygen to keep them alive.

The water quality in many of Scotland's rivers has improved over recent years. Salmon are now being caught on the River Clyde after an absence from the river of many years. The Clyde suffered badly from industrial pollution.

Quick Test

1. Hydrochloric acid was found to have polluted a loch. In a titration, 20.0 cm^3 of the loch water was neutralised by 0.1 mol 1^{-1} potassium hydroxide. The balanced equation for the reaction is:

$NaOH(aq) + HCl(aq) \rightarrow NaCl(aq) + H_2O(\ell)$

The results of three titrations are shown:

Titration	1st level (cm^3)	2nd level (cm^3)	Volume of sodium hydroxide added (cm^3)
1st	0.2	15.6	15.4
2nd	15.4	30.3	14.9
3rd	30.3	45.0	14.7

(a) Work out the average volume of sodium hydroxide used to neutralise the acid.

(b) Use your answer to (a) to calculate the concentration of the acid.

(c) What needs to be added to the acid so that you can tell at which point in the titration neutralisation has occurred?

2. Give two examples of how titrations are used in chemical analysis in everyday life.

Learning checklist

In this chapter you have learned:

Metals from ores

- metals can be extracted from their ores by heating, smelting and electrical methods, depending on their reactivity
- extraction of a metal from its ore is called reduction
- reduction of ores involves metal ions gaining electrons
- the reducing agent can be identified in a reaction

Properties and reactions of metals

- in metals the outer electrons are delocalised
- metallic bonding involves the attraction of metal ions for delocalised outer electrons of neighbouring atoms
- the properties of metals including electrical conductivity are explained by metallic bonding
- oxidation is the opposite of reduction and involves the loss of electrons by a reactant
- reduction and oxidation always take place together
- the combined reduction and oxidation reaction is called a redox reaction
- ion-electron equations can be written to describe the processes of reduction and oxidation
- ion-electron equations are found in the SQA data booklet
- redox equations can be formed by combining the ion-electron equations for reduction and oxidation
- the reactions of metals with water, oxygen and acids are examples of redox reactions and can be described using ionic equations and redox equations

Electrochemical cells

- the reactions in electrochemical cells are examples of redox reactions
- the reactions at the electrodes in electrochemical cells can be described using ion-electron equations
- ion-electron equations for the reactions at the electrodes can be combined to give a redox equation for the cell reaction

Technologies that use redox reactions

- rechargeable batteries and fuel cells are technologies that use redox reactions
- there are environmental benefits from using fuel cells

Addition polymerisation

- how three ethene molecules join to make part of a poly(ethene) molecule
- monomers for making addition polymers must have a C=C double bond
- the structure of an addition polymer can be drawn from the structure of its monomer
- the structure of the repeating unit and the monomer can be drawn from the structure of an addition polymer
- an addition polymer can be recognised from its carbon backbone

Condensation polymerisation

- the monomers used to make condensation polymers have two functional groups, one at each end of the molecule – they are bifunctional
- a small molecule is also produced when a condensation polymer is formed
- polyesters are formed when diols react with dicarboxylic acids
- how three monomers join to make part of a polyester molecule
- the structure of a condensation polymer can be drawn given the structure of its monomers
- the structure of a monomer can be drawn from the structure of a condensation polymer
- a condensation polymer can be recognised from its backbone, which contains atoms of other elements in addition to carbon

Natural polymers

- starch, cellulose and protein are natural condensation polymers made by living things
- starch is made when glucose monomers polymerise
- given a structural formula for glucose, show how three monomers join to make part of a starch molecule
- the structure of a glucose monomer can be drawn from the structure of a starch polymer
- natural polymers are biodegradable
- bioplastics can be made from renewable biomass sources such as starch, fats and oils

Creative plastics

- chemists are continually improving the the properties of plastics to widen their use
- new plastics are being developed that have properties not usually associated with plastics, e.g. solubility, water absorption, colour changing and conductivity

Ammonia – the Haber process

- ammonia is an important feedstock for the manufacture of fertilisers
- ammonia can be made in the laboratory by heating an ammonium salt with a base
- ammonia is very soluble and forms an alkaline solution when it dissolves in water
- ammonia is made industrially by the Haber process
- in the Haber process nitrogen from the air combines with hydrogen from the petrochemical industry
- the reaction of nitrogen with hydrogen to make ammonia is reversible, and the ammonia breaks down if the temperature is too high
- the Haber process needs a temperature of 450°C, a pressure of 200 atmospheres and an iron catalyst

Nitric acid – the Ostwald process

- nitric acid is an important feedstock for the manufacture of ammonium nitrate
- nitric acid is made industrially by the Ostwald process
- in the Ostwald process ammonia and oxygen are passed over a platinum catalyst at 900°C
- the reaction is exothermic so external heat can be removed when the reaction gets started
- nitrogen monoxide is initially formed, then nitrogen dioxide, which is dissolved in water
- nitrogen dioxide is made in the air during lightning storms

Synthetic fertilisers

- ammonium nitrate is formed when ammonia and nitric acid react in a neutralisation reaction
- ammonium nitrate is used as a fertiliser
- the percentage composition, by mass, of nitrogen in ammonium nitrate and other fertilisers can be calculated
- there are potentially environmental issues with nitrates being washed out of the soil

Radioactivity and radioisotopes

- there are many unstable isotopes of elements
- unstable isotopes can become more stable by emitting radiation
- isotopes that emit radiation are known as radioactive isotopes or radioisotopes
- radioisotopes have many important industrial and medical uses
- the three types of radiation emitted from nuclei are alpha (α), beta (β) and gamma (γ)
- alpha and beta radiation changes an isotope of one element to an isotope of another element

- alpha particles are helium nuclei ($^{4}_{2}He^{2+}$), i.e they are heavy positively charged particles
- alpha particles are slow moving and have low penetration – they will only travel a few centimetres through air
- beta particles are electrons and are fast moving, i.e. they are negatively charged
- a beta particle is emitted when a neutron changes to a proton in the nucleus
- beta particles are more penetrating than alpha particles but are stopped by a thin sheet of aluminium
- gamma radiation is produced by nuclei losing energy

Nuclear equations and half-life

- nuclear equations are used to describe the transitions that produce radiations
- the mass numbers and atomic numbers of isotopes are shown in nuclear equations
- the time in which half of the nuclei of a radioisotope decay is known as the half-life
- half-lives for radioisotopes are unique and constant
- the age of materials can be dated using the half-lives of radioisotopes

Chemical analysis

- chemists have an important role in monitoring our environment
- pollution can be removed before it enters our environment

Qualitative analysis

- qualitative analysis is the name given to the group of techniques that indicate which substances are present
- qualitative analysis techniques include flame testing, precipitation and chromatography

Quantitative analysis

- quantitative analysis allows the amount of a substance present to be calculated
- titration is a quantitative analysis method that is widely used in industry
- titration calculations can be used to calculate concentrations of solutions

Glossary

addition polymer: formed when lots of small unsaturated molecules (monomers) join together

addition reaction: one of the bonds of the carbon-to-carbon double bond is broken and new atoms (or groups) join to the carbon chain

alcohols: a homologous series containing the hydroxyl functional group (–O-H)

alpha (α) particles: a helium ion ($^{4}_{2}He^{2+}$) – one of the radioactive emissions given out by unstable nuclei

ammonia: the gas made in the Haber process and has the formula $\mathbf{NH_3}$

ammonium: the ion with the formula $\mathbf{NH_4^+}$. The ammonium ion is always part of a compound, e.g. ammonium nitrate

average rate: the change in the quantity of reactant or product over time

balanced equation: the total number of atoms of each element on the left-hand side of the equation equals the total on the right

beta (β) particles: a high-energy electron ($^{0}_{-1}e$) – one of the radioactive emissions given out by unstable nuclei

bifunctional: molecules with two functional groups

biodegradable: breaks down naturally over time

bioplastics: type of plastic made from renewable sources

biopolymers: natural polymer

branched chain: a hydrocarbon structure consisting of carbon atoms covalently bonded to each other with a side group attached, e.g. methyl ($–CH_3$)

carbon dating: method that uses the amount of unstable carbon-14 (^{14}C) present in a once–living material to calculate its age

carboxyl group:

```
    O
   //
 —C
   \
    O—H
```

carboxylic acids: acids containing the carboxyl functional group

combustion: the reaction of a fuel with oxygen producing energy (exothermic)

concentration: a measure of the quantity of a substance (mass or moles) dissolved in a specific volume of water (litres): c = mass / v or c = mol / v

concordant titres: titrations are carried out until the volumes measured at the point where the reaction is complete are very close to each other. The values are averaged (ignoring the rough titration)

condensation polymer: formed when lots of monomers with two functional groups join with the formation of small molecules as well as the polymer

covalent network: a giant three-dimensional structure in which all the atoms are covalently bonded to each other

cycloalkanes: a homologous series of saturated hydrocarbon with the carbons arranged in a ring

decay curve: a graph of the intensity of the radiation emitted against time

dehydration: the removal of water from a molecule, e.g. alcohol → alkene

delocalised electrons: electrons that can move easily from one atom to another

dicarboxylic acid: a carboxylic acid with two –COOH functional groups

diol: an alcohol with two –OH functional groups

direct current (dc): gives a positive and negative electrode so products of electrolysis can be identified

discharge: what a battery does when it is producing electricity

dissociation of water: a small proportion of water molecules break down into equal concentrations of hydrogen ions and hydroxide ions

dot and cross diagram: dots and crosses are used to represent electrons in diagrams showing how atoms bond

electrolysis: breaking down an ionic compound by passing a dc current through it

electrostatic forces: attraction between oppositely charged ions

emissions: particles and rays given out by unstable nuclei

endothermic: energy is taken in from the surroundings during a chemical reaction

ester link: the bond formed when a polyester is formed

esters: a group of sweet-smelling compounds formed when carboxylic acids and alcohols are warmed together in the presence of a catalyst such as concentrated sulfuric acid

exothermic: a reaction in which energy is given out

finite: will not last for ever

flue gas: gases like sulfur dioxide and carbon dioxide commonly produced as by-products of industrial processes

fuel cell: an electrochemical cell that produces electricity by combining fuels such as hydrogen and alcohol with oxygen without burning them

functional group: an atom or group of atoms (including the C=C double bond) within molecules in a homologous series that gives them particular properties

gamma (γ) rays: high-energy radiation – one of the radioactive emissions given out by unstable nuclei

general formula: the formulae of compounds in a homologous series follow a pattern, which can be represented by a general formula, e.g. alkanes C_nH_{2n+2} where n = 1, 2, 3, etc.

gram formula mass (gfm): the formula mass of a compound measured in grams

graphite: a form of carbon often used as electrodes in an electrochemical cell because it conducts electricity and will not react with the electrolytes

group ions: ions with more than one atom, e.g. sulfate (SO_4^{2-})

Haber process: industrial preparation of ammonia

half-life ($t_{1/2}$): the time in which half of the nuclei of a radioisotope would be expected to decay

homologous series: a family of compounds that have similar chemical properties and show a gradual change in physical properties, e.g. boiling point, and can be represented by the same general formula

hydration: an addition reaction that involves adding water across a double carbon-to-carbon bond to form an alcohol

hydrogel: polymer that can absorb water

hydrogenation: an addition reaction where hydrogen adds on to a molecule across the double carbon-to-carbon bond in an alkene (unsaturated) giving the corresponding alkane (saturated)

hydroxyl group: –O-H

indicator: a chemical that changes colour at the point where a reaction is complete

ion bridge: an electrolyte that allows ions to flow between the two solutions in an electrochemical cell

ion-electron equation: equations that show the electrons gained or lost by an atom or ion. The SQA data booklet gives a list of commonly used ion-electron equations

ionic lattice: a giant structure with oppositely charged positive (metal) ions and negative (non-metal) ions

isomers: molecules with the same molecular formula but different structural formula

isotopes: atoms with the same atomic number but different mass numbers

mole: the gram formula mass of a substance

neutralisation: when an acid reacts with a base to form water

nuclear equations: a shorthand way of representing radioactive decay, e.g. $^{242}_{94}Pu \rightarrow ^{238}_{92}U + ^{4}_{2}He$

nuclide notation: a shorthand way of showing the mass number and atomic number of an atom along with the symbol of the element

ores: rocks from which metals can be obtained

Ostwald process: industrial manufacture of nitric acid

oxidation: loss of electrons

photocatalysts: catalysts that absorb light energy

polyester: polymer formed when lots of carboxylic acid and alcohol monomers join

potassium–argon dating: method of dating rocks that uses the amount of argon given off when an unstable isotope of potassium decays and is trapped in the rocks

qualitative analysis: detecting chemicals that are present in a substance

quantitative analysis: detecting how much of a chemical is present in a substance

radioactive decay: name given to the process when unstable nuclei break down

radioactivity: emissions given out by unstable nuclei

radioisotopes: another name for a radioactive isotope

rechargeable batteries: batteries that, when they go 'flat', can be recharged and reused

redox equation: reduction and oxidation ion-electron equations added together

redox reaction: reaction in which loss and gain of electrons take place

relative atomic mass (RAM): average of the masses of the isotopes of an element

repeating unit: part of the structure of a polymer that repeats the length of the structure

reversible reaction: a reaction in which products break down to form reactants as they form, often represented by the $\rightleftharpoons$ symbol

reduction: gain of electrons

reducing agent: substance that brings about reduction, i.e. supplies electrons

shared pair of electrons: when two electrons from two atoms are shared, a covalent bond is formed

smelting: reducing a metal ore by heating with carbon

spectator ions: ions that do not take part in a reaction

straight chain: a hydrocarbon structure consisting of carbon atoms covalently bonded to each other with no branches (side chains)

systematic naming: an internationally agreed method of naming compounds that is used throughout the world

titration: method of carrying out a chemical reaction by adding accurate quantities of reactants to work out the unknown concentration of one of the reactants – also known as volumetric analysis

volatility: how easy it is for a liquid to form a gas – the more volatile a liquid is, the easier it is for it to form a gas

Answers

Unit 1: Chemical changes and structure

Rates of reaction

1. 0.55 (0.6) cm s^{-1}
2. 0.012 (0.01) g s^{-1}

Nuclide notation, isotopes and RAM

1. (a) $^{79}_{35}Br$
   (b) $^{79}_{35}Br^{-}$
   (c) There is a slightly higher percentage of isotope $^{79}_{35}Br$
2. RAM = 20.19 (20.2)

Covalent bonding and shapes of molecules

1. (a)

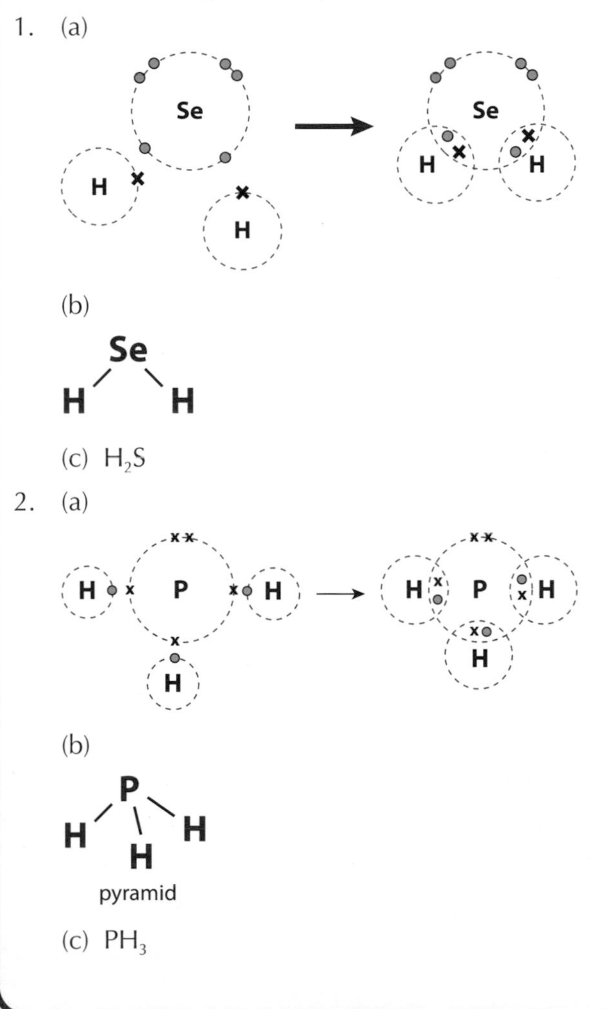

   (b) (structural formula of Se bonded to two H)
   (c) H_2S
2. (a)
   (b) pyramid
   (c) PH_3

3. (a)

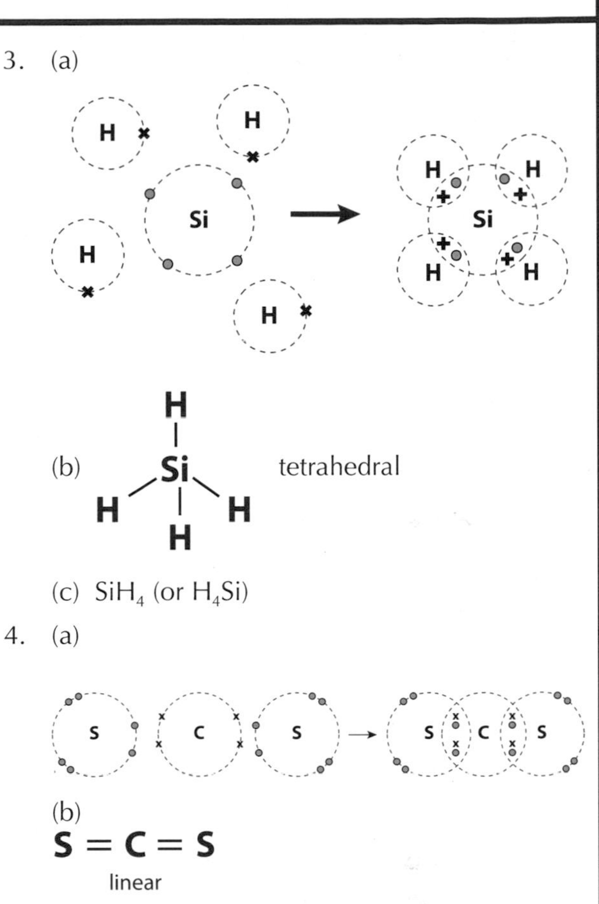

   (b) tetrahedral
   (c) SiH_4 (or H_4Si)
4. (a)
   (b) S = C = S
   linear

Structure and properties of covalent substances

(a) molecules, (b) networks, (c) gases, (d) covalent, (e) strong, (f) weak, (g) energy, (h) molecules, (i) low, (j) high, (k) covalently, (l) three-dimensional, (m) bonds, (n) insoluble, (o) dissolve, (p) electricity, (q) electrons

Structure and properties of ionic compounds

1. Ionic, because it has a high melting point and boiling point and conducts electricity as a liquid or solution.
2. His conclusion is wrong. When testing for electrical conductivity the substance has to be melted or dissolved in water. If it conducts, which magnesium chloride would, then it is ionic. Covalent substances do not conduct in any state.

3. Sodium chloride is ionic, and ionic compounds are typically solids at room temperature and have high melting points. Hydrogen chloride is covalent molecular and so has a low melting point and boiling point so is a gas at room temperature.

Chemical formulae using group Ions

1. (a) Ba
   (b) F_2
   (c) BH_3
   (d) BaO
   (e) MgS
   (f) Ca_3N_2
   (g) PCl_5
   (h) NO_2
2. (a) Na_2CO_3
   (b) $MgSO_3$
   (c) KNO_3
   (d) LiOH
3. (a) $CuCl_2$
   (b) Ag_3N
   (c) FeF_3
   (d) FeS
4. (a) $Ca(OH)_2$
   (b) $Fe_2(SO_3)_3$
   (c) $Mg_3(PO_4)_2$
   (d) $(NH_4)_2CO_3$
5. (a) $Ca^{2+}(OH^-)_2$
   (b) $(Fe^{3+})_2(SO_3^{2-})_3$
   (c) $(Mg^{2+})_3(PO_4^{3-})_2$
   (d) $(NH_4^+)_2CO_3^{2-}$

Balancing chemical equations

1. (a) $BaCl_2 + MgSO_4 \rightarrow BaSO_4 + MgCl_2$
   (b) $Fe + Cu(NO_3)_2 \rightarrow Fe(NO_3)_2 + Cu$
   (c) $CuSO_4 + NaOH \rightarrow Cu(OH)_2 + Na_2SO_4$
2. (a) $\mathbf{2}Na + S \rightarrow Na_2S$
   (b) $\mathbf{4}K + O_2 \rightarrow \mathbf{2}K_2O$
   (c) $\mathbf{2}AgNO_3 + MgCl_2 \rightarrow \mathbf{2}AgCl + Mg(NO_3)_2$

Gram formula mass and the mole

1. (a) 100 g
   (b) 108 g
   (c) 187.5 g
   (d) 132 g
2. (a) 0.075 mol
   (b) 1.5 mol
   (c) 0.1 mol
   (d) 1.75 mol
3. (a) 230 g
   (b) 21.6 g
   (c) 262.5 g
   (d) 79.2 g

Connecting moles, volume and concentration in solutions

1. 0.6 mol l^{-1}
2. 0.08 mol l^{-1}
3. 0.09 mol
4. 14 g

Acids and bases

1. (a) hydrogen, (b) equal, (c) ions, (d) non-metal, (e) below, (f) greater, (g) metal, (h) above, (i) less
2. (a) The solution is acidic so there will be a greater concentration of hydrogen ions than hydroxide ions.
   (b) As water is added the solution becomes diluted, and the concentration of hydrogen ions decreases, so the pH will rise towards pH 7.

Neutralisation and volumetric titrations

1. (a) $2H^+(aq) + SO_4^{2-}(aq) + 2K^+(aq) + 2OH^-(aq) \rightarrow 2H_2O(\ell) + 2K^+(aq) + SO_4^{2-}(aq)$
   (b) (a) $2H^+(aq) + 2OH^-(aq) \rightarrow 2H_2O(\ell)$
2. (a) burette, (b) pipette, (c) indicator, (d) colour, (e) titration, (f) neutralise

Unit 2: Nature's chemistry

Alkanes

1. (a) **Homologous series** – a family of compounds that have similar chemical properties and show a gradation of physical properties, e.g. boiling point, and can be represented by the same general formula. **Isomers** – compounds that have the same molecular formula but different structural formulae.

   (b) C_9H_{20}

2. Branching in liquid hydrocarbons causes the intermolecular forces to decrease. This means less energy is needed to separate branched molecules.

Naming and drawing branched alkanes

1. (a) 2,2–dimethylbutane

   $CH_3CH_2C(CH_3)_2CH_3$

   (b) Structural formula:

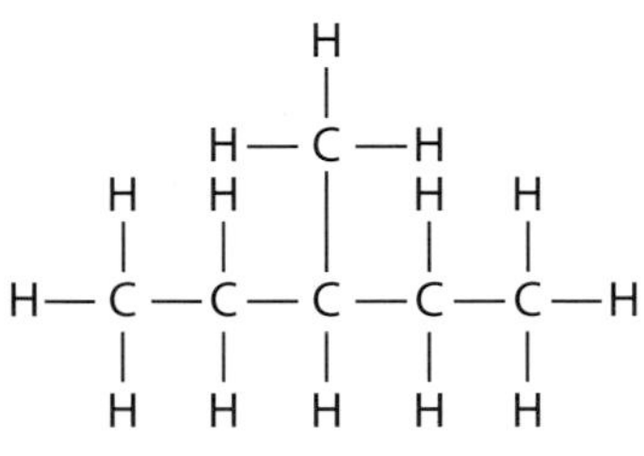

Shortened structural formula:
$CH_3CH_2CH(CH_3)CH_2CH_3$

Cycloalkanes

1. (a) C_7H_{14}

   (b) cycloheptane

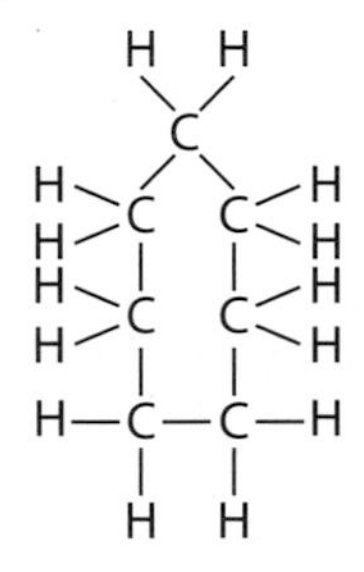

2. 

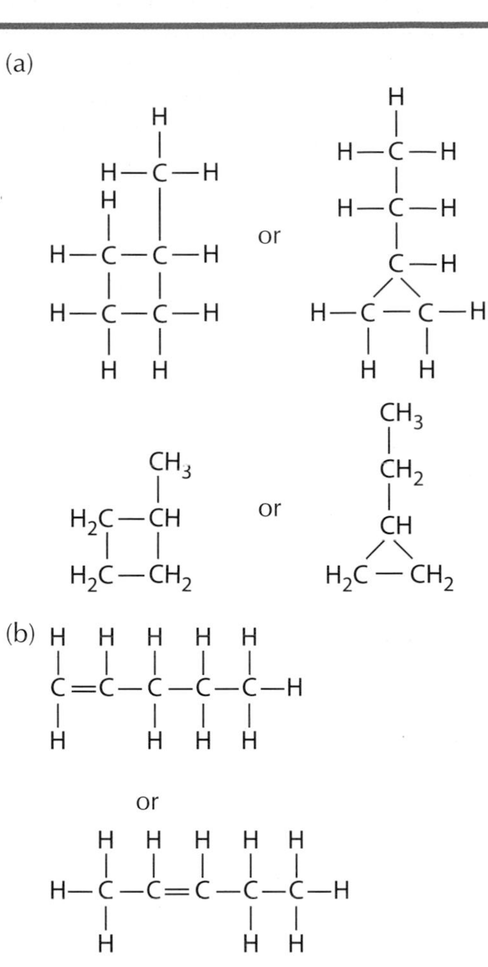

Alkenes

1. (a) A: hex–3–ene

   B: 4–methylpent–2–ene

   C: 2,3–dimethylbut–2–ene

   A: $CH_3CH_2CHCHCH_2CH_3$

   B: $CH_3CH(CH_3)CHCHCH_3$

   C: $CH_3C(CH_3)C(CH_3)CH_3$

   (b)

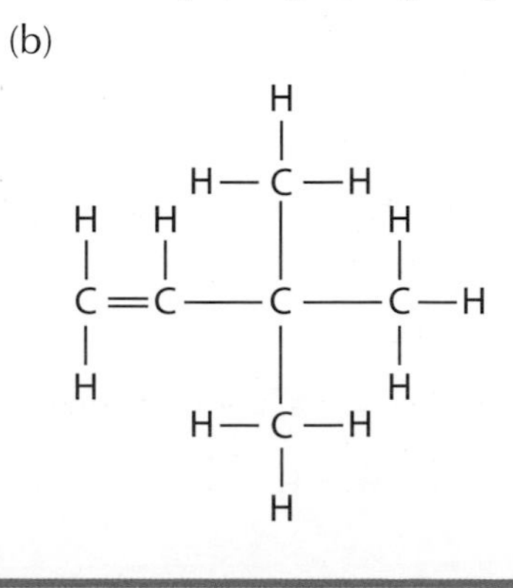

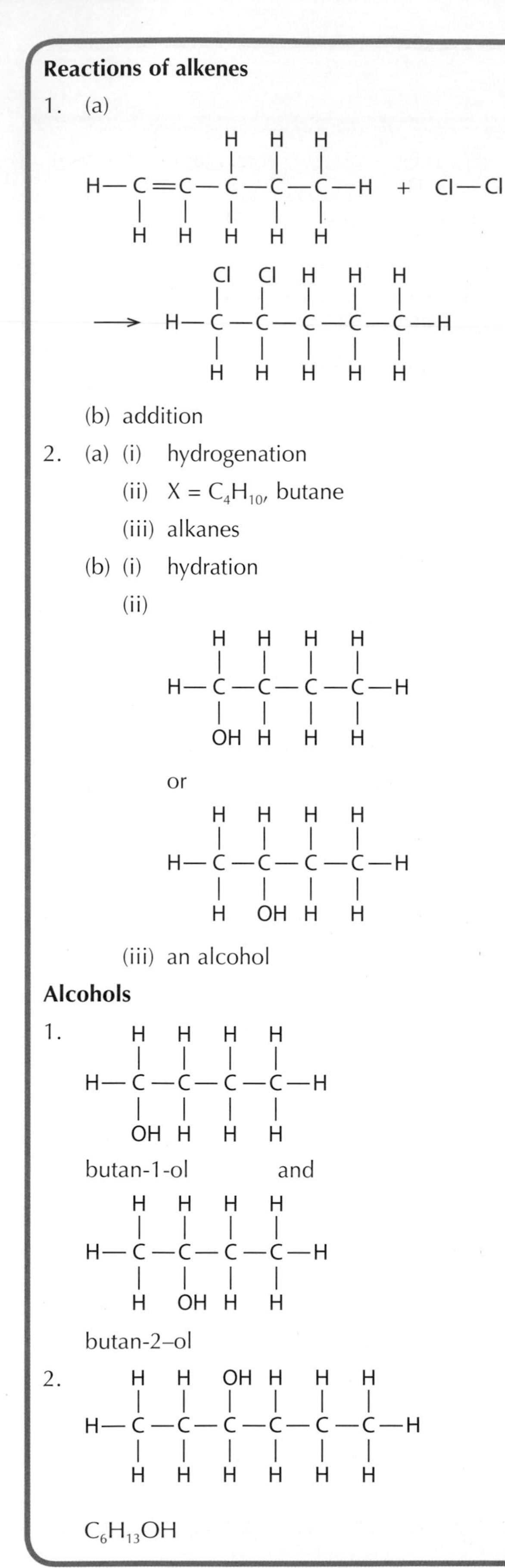

Reactions of alkenes

1. (a) $H{-}C{=}C{-}C{-}C{-}C{-}H + Cl{-}Cl \rightarrow$ $CH_2Cl{-}CHCl{-}CH_2{-}CH_2{-}CH_3$

   (b) addition

2. (a) (i) hydrogenation

   (ii) X = C_4H_{10}, butane

   (iii) alkanes

   (b) (i) hydration

   (ii) $CH_2OH{-}CH_2{-}CH_2{-}CH_3$

   or

   $CH_3{-}CHOH{-}CH_2{-}CH_3$

   (iii) an alcohol

Alcohols

1. $CH_2OH{-}CH_2{-}CH_2{-}CH_3$

   butan-1-ol and

   $CH_3{-}CHOH{-}CH_2{-}CH_3$

   butan-2–ol

2. $CH_3{-}CH_2{-}CHOH{-}CH_2{-}CH_2{-}CH_3$

   $C_6H_{13}OH$

Properties of alcohols

1. (a) Approximately 160 °C

   (b) The boiling points of the alcohols increase as the number of carbon atoms in the molecules increases. The longer the hydrocarbon chains, the greater the forces of attraction between molecules. This means more energy will be required to separate bigger molecules and so boiling points will increase.

2. For a molecule to be able to dissolve it must be able to interact with water molecules. The presence of the –OH group on the alcohol results in strong forces of attraction forming between the alcohol's –OH group and the –OH groups on the water molecules and so the small alcohol molecules mix easily with the water.

   In larger alcohol molecules, the forces of attraction between the hydrocarbon parts of the alcohol molecules are stronger than the forces between the –OH groups which makes it more difficult for large alcohol molecules to dissolve in water.

Reactions and uses of alcohols

1. (a) Ethanol burns cleanly. Ethanol can be made by fermenting sugar cane, which is a renewable energy source.

   (b) $CH_3OH(\ell) + 1\frac{1}{2} O_2(g) \rightarrow CO_2(g) + 2H_2O(\ell)$

   (c) Used as solvents

2. (a) hydration (addition)

   (b) dehydration

   (c) oxidation

Carboxylic acids and esters

1. (a) butanoic acid

   (b) C_3H_7COOH

2. (a) C_2H_5COOH

   (b)

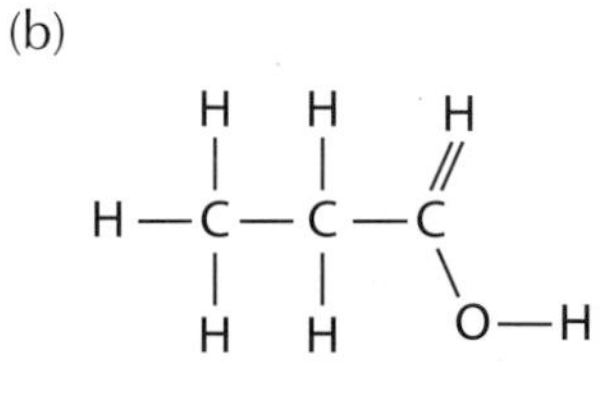

Unit 3: Chemistry in society

(c) For a molecule to be able to dissolve it must be able to interact with water molecules. Small carboxylic acid molecules are able to interact with water molecules better than larger molecules.

3. Methanoic acid and ethanoic acid. Methanoic acid is used as a preservative and antibacterial agent in farm animal feed. Vinegar is a solution of ethanoic acid in water. It is widely used as a food preservative and flavouring.

4. (a) X = carboxylic acid or alcohol, Y = alcohol or carboxylic acid, Z = concentrated sulfuric acid
   (b) catalyst
   (c) Oily layer forms on the surface of the water. Sweet smell.
   (d) Flavourings and solvents

Energy and chemicals from fuels

1. (a) $E_h = c\ m\ \Delta T = 4.18 \times 0.1 \times 33.5 =$ **14.0 kJ**
   (b) Heat loss to the surroundings. Not all the heat is transferred from the beaker to the water.
   (c) This shield cuts down on heat loss to the air. Copper is a very good conductor of heat so passes heat on to the water.
2. 95.6 g

Metals from ores

1. 86.2%
2. (a) reduction
   (b) $Cu^{2+}(s) + 2e^- \rightarrow Cu(s)$
   (c) carbon monoxide (CO)

Properties and reactions of metals

1. Metals have delocalised electrons that are not restricted to one particular atom and can move throughout the metal structure.
2. (a) $2Zn(s) + O_2(g) \rightarrow 2Zn^{2+}O^{2-}(s)$
   (b) $Zn(s) \rightarrow Zn^{2+}(s) + 2e^-$
   (c) $O_2(g) + 4\ e^- \rightarrow 2O^{2-}\ (g)$
   (d) $2Zn(s) + O_2(g) \rightarrow 2Zn^{2+}O^{2-}\ (s)$
3. (a) $2Li(s) + 2H_2O(\ell) \rightarrow 2Li^+(aq) + 2OH^-(aq) + H_2(g)$
   (b) $Li(s) \rightarrow Li^+(aq) + e^-$
   (c) $2H_2O(\ell) + 2e^- \rightarrow H_2(g) + 2OH^-(aq)$
   (d) $2Li(s) + 2H_2O(\ell) \rightarrow 2\ Li^+(aq) + 2OH^-(aq) + H_2(g)$
4. (a) $Mg(s) + 2H^+(aq) + SO_4^{2-}(aq) \rightarrow Mg^{2+}(aq) + SO_4^{2-}(aq) + H_2(g)$
   (b) $Mg(s) \rightarrow Mg^{2+}(aq) + 2e^-$
   (c) $2H^+(aq) + 2e^- \rightarrow H_2(g)$
   (d) $Mg(s) + 2H^+(aq) \rightarrow Mg^{2+}(aq) + H_2(g)$

Electrochemical cells

1. (a) $Zn(s) \rightarrow Zn^{2+}(aq) + 2e^-$
   (b) $Ag^+(aq) + e^- \rightarrow Ag(s)$
   (c) reduction
   (d) $Zn(s) + 2Ag^+(aq) \rightarrow Zn^{2+}(aq) + 2Ag(s)$
   (e) From zinc to silver, through the wire.
   (f) Ion bridge allows ions to move between beakers.
2. (a) $2I^-(aq) \rightarrow I_2(aq) + 2e^-$
   (b) oxidation
   (c) $Cl_2(aq) + 2e^- \rightarrow 2Cl^-(aq)$
   (d) reduction
   (e) $2I^-(aq) + Cl_2(aq) \rightarrow I_2(aq) + 2Cl^-(aq)$

Technologies that use redox reactions

1. (a) oxidation
   (b) Carbon dioxide, which contributes to global warming, is produced.
2. (a) $NiOOH + H_2O + e^- \rightarrow Ni(OH)_2 + OH^-$
   (b) $Cd + 2NiOOH + 2H_2O \rightarrow Cd(OH)_2 + 2Ni(OH)_2$
   (c) $Cd(OH)_2 + 2e^- \rightarrow Cd + 2OH^-$ reduction
   $Ni(OH)_2 + OH^- \rightarrow NiOOH + H_2O + e^-$ oxidation

Addition polymerisation

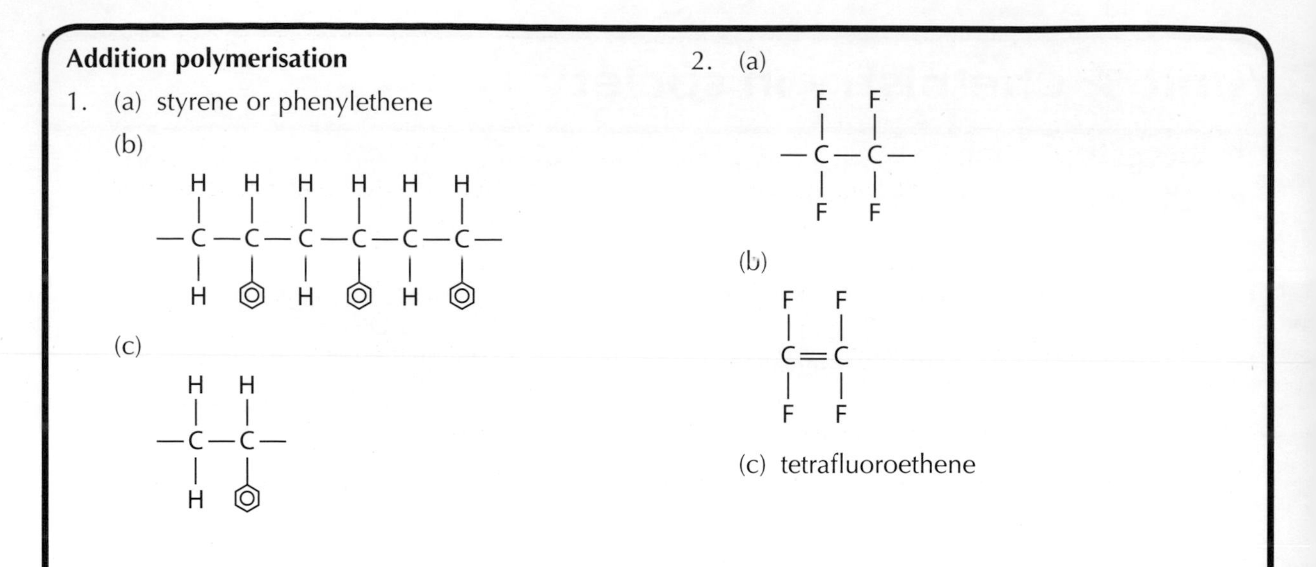

1. (a) styrene or phenylethene
   (b)
   (c)
2. (a)
   (b)
   (c) tetrafluoroethene

Condensation polymerisation

1. (a) (i) and (ii) (water)

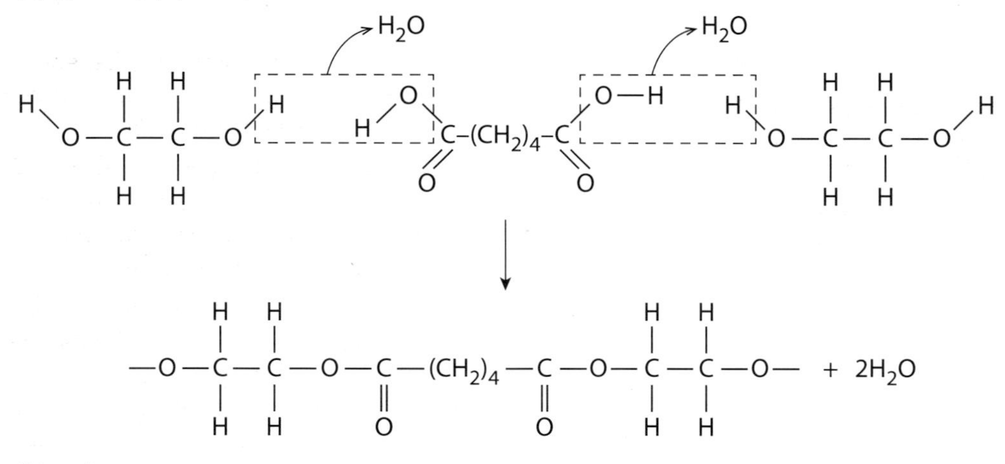

   (b) polyester

Natural polymers

1. (a) hydroxyl (–OH) and carboxyl (–COOH)
   (b) bifunctional
   (c) polyester
2. (a) natural, (b) condensation, (c) biopolymers, (d) glucose, (e) water, (f) monomer, (g) bioplastics

Creative plastics

Name/type	Property	Use
Kevlar	Strong, lightweight	Vehicle tyres, bullet-resistant vests, kayaks
Polyvinyl alcohol (PVA)	Soluble, strong	Hospital laundry bags, covering on dishwasher tablets
Sodium polyacrylate (hydrogel)	Absorbs water	Disposable nappies
RhinoPlex	Self-sealing	Instant tyre repair
Self-healing	Self-repairing when cut or damaged	Repairing cracked laptop/phone casing
Colour-changing	Changes colour when put under stress or high temperature	Incorporated into aircraft wings or bridges to warn if a crack appears
Conductive	Conducts electricity	Flexible touch screens, e-paper

Ammonia

1. (a) Mix with a base such as calcium hydroxide or sodium hydroxide or soda lime and heat.
   (b) Ammonia turns moist pH paper blue – ammonia is an alkaline gas.
   (c) Ammonia is very soluble in water.
2. (a) $N_2(g) + 3H_2(g) \rightleftharpoons 2NH_3(g)$
   (b) Around 200 atmospheres and 450°C. Iron catalyst. Recycle unreacted gases.
   (c) The reaction is reversible so as the ammonia forms some of it breaks down into reactants.
   (d) The rate at which ammonia is formed would be too slow to make it economical.
   (e) Saves valuable resources and energy, helps increase the yield of ammonia and also prevents hydrogen being released into the atmosphere.
   (f) Finely divided means there is a large surface area, which speeds up the reaction.

Nitric acid – the Ostwald process

1. (a) (a) air, (b) platinum, (c) heat, (d) ammonia, (e) nitrogen dioxide
   (b) Turns moist pH paper red.
   (c) Wire gives a larger surface area so the reaction is faster.
   (d) The reaction is exothermic, i.e. energy is given out so no external source of heat needed once reaction starts.
   (e) **Similarities:** air is used as a source of oxygen, platinum catalyst used, catalyst heated – heat eventually removed because reaction exothermic.

   Differences: in the Ostwald process: the ammonia used is produced by the Haber process, gases compressed, layers of platinum gauze catalyst used, when the nitrogen dioxide is formed it is mixed with air and passed up a tower, where it dissolves in water to form nitric acid.

Synthetic fertilisers

1. (a) ammonia + nitric acid → ammonium nitrate

   $NH_3(g) + HNO_3(aq) \rightarrow NH_4NO_3(aq)$

   (b) The water is allowed to evaporate from the solution so that solid ammonium nitrate is left behind.
2. (a) ammonia + phosphoric acid → ammonium phosphate

   $2NH_3 + H_3PO_4 \rightarrow (NH_4)_3PO_4$

   (b) Neutralisation
   (c) Soluble and contains two elements (essential elements) plants need for healthy growth.

3. $\frac{78}{174} \times 100 = 44.8\%$

Radioactivity and radioisotopes

1. (a) nuclei, (b) alpha, (c) gamma, (d) helium, (e) electrons, (f) slow, (g) beta, (h) aluminium, (i) smoke, (j) gauging, (k) cancer, (l) damage

Nuclear equations and half-life

1. (a) $^{234}_{92}U \rightarrow ^{230}_{90}Th + ^{4}_{2}He$
   (b) $^{228}_{89}Ac \rightarrow ^{228}_{90}Th + ^{0}_{-1}e$
   (c) $^{210}_{83}Bi \rightarrow ^{210}_{84}Po + ^{0}_{-1}e$
   (d) $^{45}_{21}Sc + ^{1}_{0}n \rightarrow ^{42}_{19}K + ^{4}_{2}He$
2. (a) 28 years
   (b) 46 (±1) years
   (c) 1 g
3. (a) 11 460 years
   (b) All living things contain a small amount of unstable ^{14}C and stable ^{12}C. The ratio of ^{14}C:^{12}C in a living organism remains constant throughout its lifetime as it is continually taking in carbon dioxide. When an organism dies the amount of ^{14}C in the organism begins to decrease as ^{14}C atoms decay but the stable ^{12}C does not. By measuring the ^{14}C:^{12}C ratio in the material and comparing it to the current ratio in living material, the time since the sample died can be measured.

Chemical analysis

1. The air we breathe and the water we drink are two of our essential needs. They need to be monitored to ensure they are not polluted and are fit for human use; similarly with the food we eat. Chemists ensure they are pure and will not cause us harm.
2. (a) The main air pollution problem nowadays comes from the increasing number of vehicles on our roads. The pollutants include carbon monoxide, oxides of nitrogen, unburnt hydrocarbons and tiny solid particles.
   (b) Air quality is automatically monitored at hundreds of sites across the UK or NO_2 collected in special tubes and analysed.
   (c) Sulfur dioxide is an acidic gas. It can be scrubbed by passing it through an alkaline solution – a neutralisation reaction takes place.
   (d) Photocatalysts absorb light, which helps break down pollutant gases, e.g. coatings for concrete structures and self-cleaning glass and, perhaps in the future, clothing.

Qualitative analysis

1. (a) Flame test – potassium gives a lilac flame.
   (b) Add sodium hydroxide solution – a rust red precipitate of iron(III) hydroxide would be formed.
   (c) Add silver nitrate solution – a white/cream/yellow precipitate would form if halide ions were present.
2. (a) Two
   (b) Glucose and fructose

Quantitative analysis

1. (a) 14.8 cm^3
   (b) 0.074 (0.07) mol l^{-1}
   (c) An indicator
2. In wineries titration is used to check the acidity of the wine, as this affects its keeping quality. In the dairy industry titration is part of a procedure that measures the protein content of foods. Titration allows the concentration of dissolved oxygen in water to be worked out. The oxygen level is important for the survival of fish in the water.